URBAN DESIGN
THE BASICS

Urban Design: The Basics provides a brief but compelling overview and introduction to the theory and practice of the multi-disciplinary field of urban design.

It is an 'easy-to-understand' and 'jargon-free' introduction to the fundamental principles of urban design. By introducing the essentials of urban design, the book is an important starting point for future study of the discipline. Topics include placemaking, sustainable urbanism, the evolution of cities and townscapes, and urban design and governance. Across seven chapters, the book is centred on a holistic understanding of sustainability and the special role of urban design in achieving a high quality of urban life, economic diversification, and less energy consumption. It provides a clear overview of the evolution of urban design, drawing on fundamental principles and critical challenges, and negotiates the complexities and nuances of the discipline. Given the contemporary international importance of urban design, the book uses examples from around the globe to explain its role and impact in different contexts. It also features detailed further reading lists for those wishing to expand their knowledge and understanding further.

Urban Design: The Basics will be of keen interest to those fascinated with cities and urban design, students, and practitioners who are looking to supplement their knowledge of key urban design principles.

Tim Heath is the Chair of Architecture & Urban Design at the University of Nottingham. He is a qualified architect and urban planner with over 30 years of experience in academia. He is the director of postgraduate studies in architecture and built environment and course director of the MArch in Sustainable Urban Design. He has successfully completed many research projects and has published extensively across the world in academic and professional publications. He has co-authored a number of books, written chapters in many edited books, and published extensively in refereed journals and conferences.

Florian Wiedmann is an Assistant Professor in Architecture and Urban Design at the University of Nottingham. He is an experienced urban designer with academic and professional experience in the UK, Germany, and the Middle East.

THE BASICS

The Basics is a highly successful series of accessible guidebooks which provide an overview of the fundamental principles of a subject area in a jargon-free and undaunting format.

Intended for students approaching a subject for the first time, the books both introduce the essentials of a subject and provide an ideal springboard for further study. With over 50 titles spanning subjects from artificial intelligence (AI) to women's studies, *The Basics* are an ideal starting point for students seeking to understand a subject area.

Each text comes with recommendations for further study and gradually introduces the complexities and nuances within a subject.

For more information about this series, please visit: www.routledge.com/The-Basics/book-series/B

POETRY
JEFFREY WAINWRIGHT

POLITICS AND RELIGION
JEFFREY HAYNES

SEMIOTICS (FIFTH EDITION)
DANIEL CHANDLER

EDUCATION (SECOND EDITION)
KAY WOOD

SPOKEN ENGLISH
MICHAEL MCCARTHY AND STEVE WALSH

BUSINESS START-UP
ALEXANDRINA PAUCEANU

ACTING HEIGHTENED TEXT
CATHERINE WEIDNER

LIBERTARIANISM
JESSICA FLANIGAN AND CHRISTOPHER FREIMAN

CLOSE READING (SECOND EDITION)
DAVID GREENHAM

FEMINISM
RENEE HEBERLE

MINDFULNESS
SOPHIE SANSOM, DAVID SHANNON, AND TARAVAJRA

URBAN DESIGN
TIM HEATH AND FLORIAN WIEDMANN

URBAN DESIGN

THE BASICS

Tim Heath and Florian Wiedmann

LONDON AND NEW YORK

Designed cover image: Photograph by Tim Heath and Florian Wiedmann

First published 2026
by Routledge
4 Park Square, Milton Park, Abingdon, Oxon OX14 4RN

and by Routledge
605 Third Avenue, New York, NY 10158

Routledge is an imprint of the Taylor & Francis Group, an informa business

© 2026 Tim Heath and Florian Wiedmann

The right of Tim Heath and Florian Wiedmann to be identified as authors of this work has been asserted in accordance with sections 77 and 78 of the Copyright, Designs and Patents Act 1988.

All rights reserved. No part of this book may be reprinted or reproduced or utilised in any form or by any electronic, mechanical, or other means, now known or hereafter invented, including photocopying and recording, or in any information storage or retrieval system, without permission in writing from the publishers.

Trademark notice: Product or corporate names may be trademarks or registered trademarks, and are used only for identification and explanation without intent to infringe.

British Library Cataloguing-in-Publication Data
A catalogue record for this book is available from the British Library

Library of Congress Cataloging-in-Publication Data
Names: Heath, Tim, 1964– author | Wiedmann, Florian author
Title: Urban design : the basics / Tim Heath and Florian Wiedmann.
Description: Abingdon, Oxon : Routledge, 2025. | Series: The basics |
Includes bibliographical references and index.
Identifiers: LCCN 2025027382 (print) | LCCN 2025027383 (ebook) |
ISBN 9781032161640 hardback | ISBN 9781032169750 paperback |
ISBN 9781003251200 ebook
Subjects: LCSH: City planning
Classification: LCC NA9031 .H335 2025 (print) | LCC NA9031 (ebook)
LC record available at https://lccn.loc.gov/2025027382
LC ebook record available at https://lccn.loc.gov/2025027383

ISBN: 9781032161648 (hbk)
ISBN: 9781032169750 (pbk)
ISBN: 9781003251200 (ebk)

DOI: 10.4324/9781003251200

Typeset in Times New Roman
by Newgen Publishing UK

CONTENTS

List of Figures		vi
Preface		xv
Acknowledgements		xvii

1	Urban Design: The Art and Science of Creating Successful Places for People	1
2	Evolution of Cities and Townscapes	15
3	Placemaking from Theory to Practice	41
4	Components and Principles of Urban Design	64
5	Theories on Sustainable Urbanism	141
6	Urban Design and Governance	176
7	Urban Design: Creating a Holistically Healthy City	209

Bibliography	216
Index	237

FIGURES

1.1 Successful urban spaces embrace many shapes, sizes, and types with many existing for hundreds of years (Piazza del Campo, Sienna, Italy) — 2

1.2 Cities around the world are facing considerable challenges including urbanisation, urban sprawl, and mono-functional development (typical urban sprawl of residential development in China) — 4

1.3 Many large cities have been planned to facilitate private cars at the expense of the pedestrian environment (East 2nd Ring Road, Beijing, China) — 5

1.4 Cities are also experiencing challenges from considerable traffic congestion, noise, and pollution (Tainan City, Taiwan) — 6

1.5 Urban design is fundamental in the creation of safe, inclusive, and enjoyable public places (Old Market Square, Nottingham, UK) — 9

4.1 Successful public streets create pedestrian-friendly environments that are inclusive and provide many different reasons for people to use them (Qianmen Street, Beijing, China) — 66

4.2 Successful public squares attract a wide range of users for a variety of different activities (Kultorvet Square, Copenhagen, Denmark) — 67

4.3	Streets can serve many different functions that contribute to their character and qualities (Jalan Hang Kasturi in Kuala Lumpur, Malaysia)	70
4.4	The activities on a street together with the enclosure created by the building facades help to create its character (Rue Nicolas Flamel, Paris)	71
4.5	Streets can be formal and grand in their appearance resulting from the layout and surrounding buildings (Rua Augusta, Lisbon, Portugal)	72
4.6	Streets can be more informal and organic in their layout and character (Steep Hill, Lincoln, UK)	73
4.7	Public squares can be intimate and informal in shape, scale, and appearance (Neal's Yard, London, UK)	75
4.8	Public squares can be formal and grand in appearance whilst also offering a strong sense of enclosure (Place des Vosges, Paris, France)	76
4.9	Public squares can come in many different shapes and sizes (Plaza Redonda, Valencia, Spain)	77
4.10	Squares that are uncomfortable in scale and have very few reasons for people to use them tend to be less successful (City Hall Plaza, Boston, USA)	77
4.11	Successful public squares often have the flexibility to accommodate different activities and events at different times (Old Market Square, Nottingham, UK)	78
4.12	Public squares should provide reasons for people to use them at different times of the day and in different seasons of the year (Plaça de la Independència, Girona, Spain)	79
4.13	Context is a primary consideration in urban design and the public square as a formal marketplace in common within many cultures (Old Market Square, Nottingham, UK)	83

4.14 Public spaces can also accommodate more informal markets such as night markets (Garden Night Market, Tainan City, Taiwan) — 84

4.15 In some contexts, urban squares can have more of a ceremonial role as a celebration of power or as a place of display (Tiananmen Square, Beijing, China) — 86

4.16 Markets are great displays of local culture and in some countries the grand market hall is a dominant urban feature (Central Market, Valencia, Spain) — 86

4.17 Streets with inactive facades providing no natural surveillance create an unattractive and unwelcoming environment that often convey a perception of being unsafe — 92

4.18 Streets that have active edges with high levels of natural surveillance tend to be more attractive and feel safe (Huangshan, Anhui Province, China) — 93

4.19 Ground floor functions that spill-out into the space help to enliven streets, provide opportunities for people watching, and enhance activity (Rue Soufflot, Paris, France) — 94

4.20 Shops and cafes spilling out onto the street provide both visual interest and increase natural surveillance (Ruga dei Oresi, Venice, Italy) — 95

4.21 Providing vistas of well-known landmarks can enhance legibility and wayfinding within cities (Calle Padre Hortas Cáliz, Jerez de la Frontera, Spain) — 97

4.22 Landmarks that terminate vistas help to induce movement through the urban environment (Yasaka Pagoda, Kyoto, Japan) — 98

4.23 Successful public spaces are inherently flexible and able to host a wide range of different events and activities (Place de l'Hôtel de Ville, Paris, France) — 99

4.24 New public spaces can be created through the adaptive re-use of brownfield sites that previously had industrial uses (Coal Drops Yard, London, UK) — 100

4.25	Abandoned transport infrastructure in cities, such as former railway lines, can also be adapted into successful public places (The High Line, New York, USA)	101
4.26	Mixed-use developments such as shophouses create many different reasons for people to visit a place and help to stimulate activity throughout the day and night (Smith Street in Singapore)	105
4.27	Neighbourhoods that contain many different functions tend to attract a wide variety of different users and encourage people to walk from one activity to the next (Dafen Village, Shenzhen, China)	106
4.28	Neighbourhoods with buildings of different size, condition, and tenure help to promote opportunities for small local businesses which can boost the local economy and also create places with unique characteristics (Tianzifeng, Shanghai, China)	109
4.29	Poor quality pavements especially with poorly located street furniture, lampposts, and signs can discourage walking and cycling and be very difficult for people that are less mobile (Kuala Lumpur, Malaysia)	112
4.30	The elderly, children, and those with physical challenges can find urban environments hostile and dangerous if they are not designed appropriately	113
4.31	Successful walkable urban environments tend to be well-maintained, safe, and welcoming places (Plaza de la Yerba, Jerez de la Frontera, Spain)	114
4.32	Different materials, textures, and colours can help to demarcate zones for different users whilst also creating pedestrian-priority public spaces (Calle Molina Lario, Malaga, Spain)	116

4.33 Successful streets can be multi-modal and accommodate pedestrians and public transport (Goldsmith Street, Nottingham, UK) — 117

4.34 Dedicated cycling infrastructure can encourage safe and convenient travel by bicycle (Cykelslangen, Copenhagen, Denmark) — 118

4.35 Dancing in public is a popular activity that attracts a diverse range of people and enlivens the street scene in some cultures (Shanghai, China) — 120

4.36 Water features can enhance urban environments and also encourage play and interaction (Taikoo Li Sanlitun, Beijing, China) — 121

4.37 Spontaneous activities such as street performers enliven spaces, create visual interest, and attract visitors (South Bank, London, UK) — 121

4.38 Steps create important places for people to rest, to eat and drink, to meet friends, and to people watch within cities (Église de la Madeliene, Paris, France) — 124

4.39 Formal and informal seats, such as low walls, can make excellent places for people to rest, meet friends, and people watch (Suzhou, China) — 125

4.40 Opportunities to sit alongside water provide an attractive location within cities (Rathaumarkt, Hamburg, Germany) — 125

4.41 Steps and seating linked to buildings can create public amphitheatres for congregating, watching events, and observing people (Sea World Culture and Arts Center, Shenzhen, China) — 126

4.42 The combination of water and greenery creates an attractive urban environment, especially in warmer climates (KLCC, Kuala Lumpur, Malaysia) — 127

4.43 Waterside places with shade from trees can provide attractive locations for cafes and restaurants (Suzhou, China) — 128

4.44	Canal networks and rivers with their distinctive characteristics create unique and attractive contexts in many towns and cities (Groenburgwal, Amsterdam, Netherlands)	129
4.45	Temporary activities help to transform public spaces and attract different users (Festival City, Dubai, United Arab Emirates)	132
4.46	Pop-up activities such as bookstores can transform underused spaces within cities (South Bank, London, UK)	133
4.47	Urban spaces that are badly designed, with no purpose, and poor maintenance discourage users and therefore feel unsafe and impact negatively upon the local environment	135
4.48	Strong management of public places can include seasonal events with temporary attractions that entice many visitors (Paris Plage, Paris, France)	135
4.49	The most successful public spaces tend to be inclusive of, cater for, and attractive to all members of society regardless of age, gender, race, etc. (Old Market Square, Nottingham, UK)	136
4.50	Rooftops, even those that are privately owned, can become seasonal social spaces and offer alternative places and views for visitors (Galeries Lafayette, Boulevard Haussman, Paris, France)	137
4.51	Places that continue to attract people at all times of day, on all days of the week, and at all times of the year tend to be well-managed to ensure that they are clean and safe with a range of varied activities to cater for all interests (Rambla de la Llibertat, Girona, Spain)	138
5.1	A gentrified neighbourhood with refurbished buildings (Notting Hill, London, UK)	147

5.2	The new transit-oriented district of Riedberg with shared green spaces in the urban periphery of Frankfurt am Main (Germany)	149
5.3	Vernacular neighbourhood streets in the historical and restored settlement of Diriyah near Riyadh (Saudi Arabia)	153
5.4	A contemporary neighbourhood street in Riyadh (Saudi Arabia)	154
5.5	Suburban housing in Tacoma, Washington State (USA)	157
5.6	Piccadilly Circus is a public square on top of a busy underground station (London, UK)	161
5.7	Apartment blocks in Juffair (Manama, Bahrain)	163
5.8	The pedestrian experience in Dubai Marina (Dubai, UAE)	169
5.9	A scenic walk exposed to nature and the historic townscape (Salzburg, Austria)	170
6.1	A typical UK city centre street showing the decline of retail activities (Nottingham, UK)	182
6.2	Cars parking at a new metro station in Riyadh (Saudi Arabia)	186
6.3	The commercial centre of Milton Keynes New Town (UK)	189
6.4	The block structure defined by nineteenth-century planning in Berlin (Germany)	193
6.5	The typical walking experience in new edge cities in Beijing (China)	195
6.6	New developments at Liverpool's historic port and waterfront (UK)	197
6.7	High-rise buildings in the City of London (UK)	198
6.8	The pedestrianised Buchanan Street in Glasgow (UK)	199

7.1	The public realm is increasingly important in promoting physical, mental, and spiritual well-being (Tainan, Taiwan)	211
7.2	The ability to play and interact in public space helps to attract people of all ages to use the public realm and helps to build a sense of community (Granary Square, London, UK)	211
7.3	Creating opportunities for exercise and play in public spaces are an important part of promoting healthy and cohesive communities (Wenxin Park, Shenzhen, China)	213

PREFACE

Urban Design: The Basics is inspired by our passion for the role that urban design can play in placemaking and the creation of resilient and sustainable people-focused towns and cities. Indeed, our intention with this book is to inform and inspire those interested in discovering more about the importance and role of urban design as a discipline and particularly those that are considering studying urban design and perhaps becoming future practitioners.

Urban design is a broad and multi-disciplinary, and therefore, *Urban Design: The Basics* intended as an introduction for those approaching a subject for the first time. As such, it provides a starting point in the form of an overview from the need for, definition of, and evolution of urban design through to the fundamental principles and critical challenges in order to gradually introduce the complexities and nuances facing the discipline. Importantly, the book is written and presented in a clear format with many links to further reading and resources to enable readers to further develop their knowledge and understanding of urban design.

Our hope is that *Urban Design: The Basics* will therefore introduce readers not just to urban design as a subject and discipline, but also to

- new ways of looking at, seeing, and experiencing the city;
- question and analyse the city;
- understand who is using the public realm together with when, how, and why;

- ask questions about what is working well and what is not in urban environments; and
- common-sense issues relating to cities and urban areas (it's not 'rocket science'!).

In doing so, we hope to inspire the next generations of urban designers and city-lovers.

Tim and Florian
May 2025

ACKNOWLEDGEMENTS

We would like to first express our utmost love and thanks to our families who have put up with us and supported us not only during the writing of this book but all of the other times that we have been distracted or overworked in our academic positions. So, a massive thank you to Karen, Marcus, and Charlotte Heath and to Phyllis, Asiyah, and Adam Wiedmann.

Importantly, we would also like to express our gratitude to the many colleagues and ex-colleagues who have shared our, sometimes enjoyable, rewarding, frustrating, and stressful, experiences of academic life: teaching, researching, and undertaking management and administrative duties.

Importantly, we would also like to share our appreciation to the thousands of undergraduate, postgraduate, and PhD students that we have had the pleasure to teach and share time and ideas during our careers. In some way, they have all contributed to this book by helping us to continue to learn through the informed discussions, debates, and questions that we have had over the years. As we always say: *"the day that we stop learning will be the day that we stop teaching."*

Finally, we'd like to thank the publishing team at Routledge for all of their efforts in the realisation of this book and in particular for continuing to push us when we were getting distracted by the many other demands of academic life.

URBAN DESIGN
The Art and Science of Creating Successful Places for People

To serve as an introduction to the subject and discipline of urban design, this chapter is organised into four main sections that will: discuss why we need urban design; describe the emergence of urban design; define what urban design is; and highlight the roles of the urban designer.

WHY WE NEED URBAN DESIGN

The need for urban design, and urban designers, has grown significantly over the past thirty years or so because of the urgent need to address a wide range of challenges and issues facing urban areas across the world. Although varying in different cities, countries, and regions, these critical situations are primarily impacted by population growth, urbanisation, climate change, and poor decision-making and design execution from previous eras (Figure 1.1). Whilst there are a lot of interrelated factors that are impacting and influencing each of these that require in-depth examination, each will be briefly discussed in this section.

The United Nations (2025) highlights how the world's population has been rapid over the past seventy years and is now *"more than three times larger than it was in the mid-twentieth century."* As a result, the global population surpassed 8 billion people in mid-November 2022, significantly 1 billion more than in 2010. Whilst the overall rate is now considered to be slowing, this dramatic

DOI: 10.4324/9781003251200-1

Figure 1.1 Successful urban spaces embrace many shapes, sizes, and types with many existing for hundreds of years (Piazza del Campo, Sienna, Italy).
Source: Photograph taken by the authors.

growth has placed an increasingly critical burden upon issues like healthcare, education, food, and housing. A significant impact from an urban design perspective has been the contribution of population growth to what was already an increasingly urbanised world. Indeed, urbanisation, or the increase in people living in towns and cities, has led to a significant expansion in the size and number of cities and megacities around the world. Indeed by 2030, it is anticipated that there will be 43 megacities with population of over 10 million inhabitants (United Nations 2024). By 2023, 56 per cent of the world's population, or 4.4 billion people, lived in cities and this is predicted to double by 2050 to over 70 per cent. Significantly, most countries in Europe already exceed over 75 per cent of people living in urban areas. These trends will therefore continue to place unprecedented pressures upon cities and all built environment professions will have a critical role to play in the attempts to create urban environments that can meet the needs and requirements of all of these inhabitants.

Concerns about climate change emerged in the mid-1970s; however, since 2000, these have been at the forefront of the world's social, economic, and political agendas. Significantly, it is a primary concern for urban areas, both in terms of its impact upon settlements but also the contribution that urban areas are having to climate change. Indeed, it is estimated that urban areas account for around 70 per cent of global CO_2 emissions with buildings and transport being the largest contributors (UNEP 2024). More sensitive and sustainable design solutions together with better management and operational decisions for cities can therefore play a crucial and significant role in helping to reduce carbon emission and in turn global warming. In addition to the impact of cities upon CO_2 emissions, urban areas are themselves being impacted upon global warming due to rising temperatures and heat stress, pollution, water security, extreme weather events, flooding, and fires. These and other factors are, therefore, all critical considerations not just for urban design but for all agencies that are responsible for the design and management of urban settlements across the world (Figures 1.2–1.4).

Many of the world's cities and urban areas have evolved over a long period of time, even over many centuries (see Chapter 2), and as a result, most contain many areas with challenges and problems inherited from previous planning, design, and management decisions. These can relate to the architecture, urban planning, transport planning, and urban management of towns and cities and have significant and negative impacts upon the people and communities that inhabit urban areas. Indeed, even the most popular and successful settlements will include areas that have issues that need addressing. Many of these will be the result of an over-reliance on land-use zoning in planning policies and the approaches to design that have enabled the motor car to dominate urban areas. The negative impacts of land-use zoning have included: mono-functional development; a lack of diversity; a lack of activity; poor access and connections; isolation; a lack of natural surveillance; placeless environments; and a lack of character. In addition, over-reliance on cars has created further issues such as: traffic congestion; pollution; urban sprawl; pedestrian safety; and poorer provision for walking, cycling, and public transport. The challenges that these approaches have created and the importance of dealing with them through urban design will be discussed further in Chapter 4.

Figure 1.2 Cities around the world are facing considerable challenges including urbanisation, urban sprawl, and mono-functional development (typical urban sprawl of residential development in China).

Source: Photograph taken by the authors.

EMERGENCE OF URBAN DESIGN

The design of urban forms has been a primary concern for those responsible for the design and creation of settlements for thousands of years (see Chapter 2). Given the focus of those responsible for shaping these projects tended to be predominantly architects, landscape architects, or urban planners, however, the focus was often predominantly on structure and order whilst contexts were often neglected. In particular, the design of the urban form and especially the public realm often failed to consider the human dimension with there being a preoccupation with the physical and aesthetic qualities of urban environments.

Figure 1.3 Many large cities have been planned to facilitate private cars at the expense of the pedestrian environment (East 2nd Ring Road, Beijing, China).

Source: Photograph taken by the authors.

Awareness of these issues led to urban design being discussed in the American planning profession in the mid-1950s and subsequently led to a series of pivotal Ivy League conferences in the United States. Significantly, the first urban design conference was hosted at Harvard University in 1956 by Josep Lluís Sert who was Dean of the Graduate School of Design. Indeed, with eminent participants including Jane Jacobs, Lewis Mumford, and Edmund Bacon, the conference marked the beginning of urban design as an intellectual discipline and as a professional direction that was distinct from architecture, urban planning, and landscape architecture.

Figure 1.4 Cities are also experiencing challenges from considerable traffic congestion, noise, and pollution (Tainan City, Taiwan).

Source: Photograph taken by the authors.

Sért organised a further series of symposia, to begin to define the emerging discipline of urban design and to shift thinking about cities to a more holistic perspective. This approach stemmed from a need to move beyond civic design to focus more upon the living conditions of people and communities that were not being adequately addressed by the architecture and planning professions. As a result, and despite some initial opposition, urban design quickly established its own position, however more as a field of activity than as a professional discipline in its own right.

Significantly, recognition for urban design was rapid, and by 1959, the American Institute of Chartered Planners recognised its importance in a policy statement on urban renewal, whilst in 1960, the American Institute of Architecture established a Committee on Urban Design. Also, from 1959, a further series of conferences on urban design had been held in the University of Pennsylvania and these provided a platform for Harvard's Graduate School of Design to develop a curriculum and launch the first university programme in urban design in 1960. This was seen as their fourth design discipline, alongside its historic programmes in architecture, landscape architecture, and town planning (Black 2024). Later in the UK in 1972, the Joint Centre for Urban Design was launched at Oxford Polytechnic, and subsequently, the UK's Urban Design Group was established in 1978. Interestingly, however, Punter and Carmona (1997) identified that the term urban design was not used in any government publications or guidance until much later in the discussion document *Quality in Town and Country* (DoE 1994).

Since its birth in the mid-to-late 1950s, theorists and designers have rapidly developed research and tested ideas that have enabled urban design to emerge into the holistic discipline that it has become by the twenty-first century. Indeed, there have been a number of highly influential academics, theorists, and practitioners whose research and practice of urban design and placemaking have significantly impacted the discipline. Indeed, much of their thinking and approaches remain critical to the position of urban design and understanding of the urban design process. This established a shared ambition of a human-centric and community-focused approach being central to the analysis and design of urban environments. Many of the outputs of these pioneering individuals and organisations have become authoritative seminal works on urban design and these will be discussed in more detail in Chapter 3.

DEFINING URBAN DESIGN

Urban design is critical to balancing the social, environmental, and economic outcomes in the design and creation of towns and cities. As such, it plays a key role in ensuring the delivery of development that is aligned to the needs of people and communities. Unlike other built environment professions, such as architecture, town planning, landscape design, and surveying, urban design is a discipline without chartered status. It does, however, have an intertwined relationship with these and others but is considered as an inclusive as opposed to exclusive discipline.

It is important to stress that urban design engages with urban areas at three fundamental scales, from the macro-level (whole city) to the meso-level (neighbourhood), and the micro-level (public space). It is also about the process (the policy, planning, conception, and design of a project) through to the product (design realisation, activation, and management of the public realm). The overriding aim of urban design, however, is to create places that are sustainable and resilient, successful socially and economically, and importantly surpass the needs and expectations of all users.

Accurately defining urban design has long been a challenge, and it feels like almost every document or publication concerning urban design attempts to deliver its own definition of what urban design means. It is the breadth of the roles of urban designers together with those engaged in the urban design process, and how its relationships with other professions, that is behind the challenge of clearly explaining exactly what urban design actually is. As such, the term urban design has been defined in many scopes, approaches, and classifications with varied scales and interpretations. Whilst most of these definitions are more or less accurate, there are as many that are inappropriate or overly simplify what urban design is, which does not help with the wider understanding of what is a relatively new discipline (see the section on the Emergence of Urban Design). Indeed, given the multi- and inter-disciplinary nature of urban design together with its broad scope of activities and interests, the definition of urban design has been endlessly debated. In this book, however, we define urban design as *the art and science of creating successful places for people.*

This definition is inspired by many others including the fuller definition in the UK government's report By Design (DETR 2000). In this document, urban design is described as including *"...the way places work and matters such as community safety, as well as how they look. It concerns the connections between people and places, movement and urban form, nature and the built fabric, and the processes for ensuring successful villages, towns and cities"* (DETR 2000: 8). Significantly, this includes not just the activity of design but also how to achieve urban design with the process being as important as the practice. As such, it emphasises the essential roles of function or use, performance, appearance, pragmatism, accessibility and inclusivity, and, significantly, how to make these happen (Figure 1.5).

Urban design is a multifaceted and intricate concept, which the Urban Design Group (2025) states *"...is a collaborative and multi-disciplinary process of shaping the physical setting for life – the*

Figure 1.5 Urban design is fundamental in the creation of safe, inclusive, and enjoyable public places (Old Market Square, Nottingham, UK).

Source: Photograph taken by the authors.

art of making places …[it] involves the design of buildings, groups of buildings, spaces and landscapes, and establishing frameworks and procedures that will deliver successful development by different people over time." Indeed, urban design is a problem-solving process that can significantly influence the economic, environmental, social, and cultural outcomes of a place. It is concerned with analysing, organising, and shaping urban environments to *"inspire, illustrate, and define how a place can be improved or protected to bring benefits to investors, developers and wider society"* (Urban Design Group 2025). Indeed, although urban design is often delivered as a specific project, it is in fact a long-term process that continues to evolve over time.

In the latter part of the twentieth century, urban design became *"a mainstream professional activity [that is more than another built environment profession: it is a set of skills, a state of mind and a way of thinking"* (Urban Design Group 2024). As such, urban designers place greater emphasis on the needs of the public and communities in coordinating the relationships between people, communities, and places. Urban design projects should be people-focused and contextually sensitive whereby the 'real' clients should be the public rather than the individual or organisation that might have commissioned the project.

ROLE OF THE URBAN DESIGNER

As mentioned in previous sections, urban design, whilst being its own discipline and area of professional activity, has close synergies with architecture, urban planning, and landscape design. As a result, many professional companies that specialise in these areas also have expertise in urban design and many professionals will have qualifications and skills in more than one of these. It is important, however, to emphasise that urban design also has much broader overlaps with many other professions, disciplines, and areas of expertise, such as real estate, civil engineering, transport planning, conservation, sociology, psychology, law, politics, economics, human geography, natural sciences, and tourism.

Indeed, urban design is a far-reaching inter-disciplinary and multi-disciplinary concept that engages with and requires input from a wide range of professions and disciplines. Most importantly, it requires participation from all actors, or players, involved in the design, use, management, and maintenance of the built environments that are being considered. It is essential, therefore, that all of these interested parties are enabled to play a part in the urban design process. It is this reason that makes it essential for urban design to be an inclusive discipline that welcomes wide and meaningful engagement as opposed to be an exclusive and restricted profession. Although the urban design process, as explained, requires contributions from a plethora of key players, the fundamental role of the urban designer is essential in achieving successful urban solutions. Indeed, it is imperative that the urban designer can bring together the ideas and knowledge from others to resolve challenges and create better places for everyone. The Urban Design Group (2025) note how when achieved successfully, this can result in new places being constructed or a new appreciation of existing urban areas.

The role of the urban designer itself is also very broad and therefore requires many different skill sets, areas of expertise, and varied knowledge. Indeed, one of the attractions of urban design as a professional activity is that it can provide important roles for individuals from many different academic and disciplinary backgrounds. As such, those progressing into urban design can obviously come from the closely related disciplines of architecture, urban planning, and landscape design, but also from a wide range of others including geography, law, sociology, and transport. All of these backgrounds can bring essential expertise to the many roles of urban designers, and these become more apparent when considering the range of tasks that urban designers engage with. These roles span from

- contributions to policy, strategy, and plan-making; preparing design guidance, design briefs, and development frameworks;
- undertaking research, information collection, and data analysis;
- creating vision statements and design concepts;
- preparing feasibility studies;
- consultation and engagement with the public and other key actors;
- developing site-specific masterplans and land-use planning;
- designing the public realm and urban blocks;
- preparing design illustrations and other presentational material;
- liaising with and coordinating other professional consultants;
- contributing to activity and event planning; and
- preparing on-going management and maintenance manuals.

These roles are not exhaustive or explained in detail here; however, they highlight the wide spectrum of skills required and opportunities within urban design practice. In different situations, each of these might also require a different level of detail in terms of the resolution with some projects only being developed to an indicative level and others being more specific, or some requiring strategic design and other being developed to a detailed level. Importantly, the urban design process itself also extends beyond this into roles like town centre management, activity and event planning for urban spaces, and transportation management. Importantly, it is only when all of these are employed synergistically that successful, inclusive, enjoyable, and safe public places and urban environments are likely to be achieved.

Employment opportunities for urban designers are also quite diverse. Indeed, whilst a few urban designers work as sole practitioners due to the breadth of skills required some might work as individual consultants for larger companies, authorities, or organisations. Many will be employed as part of larger teams by urban design companies and especially by multi-disciplinary practices or companies specialising in built environment disciplines. Also, others will work directly for organisations that have significant land holdings such as developers, or others with interests in the land and real estate development. Meanwhile, the public sector also employs many urban designers within local planning authorities or local governments, and a smaller number in more advisory roles with central government departments. In most of these roles, the urban designer will also be engaging with a wide range of clients such as real estate developers, local government authorities, community groups, and other organisations (Urban Design Group 2025).

SUMMARY

Urban design is a multifaceted discipline that plays a fundamental role in the planning, designing, and adaptation of cities, towns, and other urban areas. The urban design process aims to create places that are sustainable, resilient, functional, and visually beautiful for the people who live and use urban environments. As such, urban design needs to adopt people-focused, human-centric approaches that consider the needs, preferences, and requirements of everyone who lives, works, plays in, and visits urban spaces. Achieving this is essential in helping to determine the quality of life in neighbourhoods, towns, and cities.

STRUCTURE OF THE BOOK

This book is structured into seven chapters that will introduce the reader to the basics of urban design and hopefully inspire further reading and exploration of urban design as a discipline, and potentially as an academic subject, and future career.

Chapter 1 has delivered a short introduction to why urban design has become increasingly important in contemporary towns and cities. It has also discussed the emergence of urban design, defined what urban design means, and highlighted the role of the urban designer.

Chapter 2 traces the evolution of cities and townscapes by focusing on significant periods in terms of approach, growth, and change in order to explain how existing cities have emerged and been influenced by historical precedents. It also covers major recent concepts for what the contemporary city should be if it is to serve the needs of its people.

Chapter 3 focuses upon the evolution of placemaking from theory to practice by examining pioneering academics, theorists, practitioners, and organisations who have strongly influenced how we undertake urban design in the twenty-first century.

Chapter 4 examines what makes successful public places and then considers the fundamental physical components of urban design together with the impact of global contexts upon the discipline. Subsequently, the chapter explores the various key principles that need to be considered during the urban design process by looking at important criteria under the physical, functional, social, and operational principles. This chapter contains many photographs of original and unique examples to illustrate the urban design principles and components discussed; however, readers should also refer to examples and diagrams in the recommended further reading.

Chapter 5 explores the theory of sustainable urbanism by first reviewing the scientific discipline and its historical roots. Since the first half of twentieth century, the Chicago School of Sociology has been an important factor in establishing the field by investigating both social groups and their spatial practices. In the second half of the twentieth century, Henri Lefebvre established his theory on space production concluding with the vision of overcoming abstract systems of decision-making. The chapter concludes with a discourse on the two main dimensions enabling more efficient ways of urban living: spatial diversification and a shared sense of belonging.

Chapter 6 considers how urban design is managed as part of urban governance and how different project types can be distinguished. It also explores how urban design provides a service centred on forming everyday urbanism as part of the process of urban governance and raises questions about who is actually responsible for the city.

Chapter 7 summarises the content of the previous chapters but also focuses upon emerging challenges and opportunities whilst briefly speculating on the future direction of the discipline of urban design.

FURTHER READINGS

This book, by its very nature, is an introductory text on urban design that the authors hope will inspire further investigation. Each chapter (and many sections) in the book will, therefore, guide the reader towards recommended further readings. This will enable a more in-depth understanding of the topics and their influences upon urban design, urban form, the public realm, and the lives of people in contemporary cities. Some of the publications recommended further readings are classic texts and these have often been reprinted or republished so, where possible, please refer to the latest editions. Also, some may be difficult to obtain due to their age; nevertheless, good libraries should have copies available. Many of these key texts have also been translated into various international languages which will assist some readers.

EVOLUTION OF CITIES AND TOWNSCAPES

Cities are complex organisms that are in a continual state of change. Human settlements have existed for thousands of years since civilisations started to become more organised usually for trading purposes and security. As a result, urban forms started to appear that displayed a logic in terms of activities, connections, spaces, and structures. Significantly, from the earliest known settlements onwards, cities began to form part of a wider urban system with their own social hierarchies, power, and wealth.

In terms of the urban forms created, there are typically two types of cities: those that have been planned with an organised urban layout and those that are unplanned and display a more organic pattern. As many cities have evolved and experienced various levels of growth during different eras, however, they can typically display areas of both formal structure and order together with others that are informal and more organic in layout.

This chapter will trace the historical evolution of towns and cities through the approaches to urbanism and design adopted in their creation and adaptation. The design of cities will be examined in order to understand the evolution and origins of design ideas in the contemporary city. This will chronologically commence with the earliest planning of settlements in ancient civilisations, through the significant development of urban forms by the Greeks and Romans, the grandeur of the Renaissance, the ideals of the Beautiful City, the rationality of Modernist urbanism, through to more human-oriented approaches in the late twentieth and early twenty-first centuries.

ANCIENT CITY PLANNING

The design, planning, and creation of world's first major cities can be traced back thousands of years to ancient civilisations. Indeed, larger organised settlements were needed as people began to cultivate crops, live in communities, and trade on a larger and more coordinated scale (Morris 1994). An urban revolution was taking place, and this process of urbanisation was accompanied by trade on a new scale. The first recognised cities were created on fertile lands in the historic region known as Mesopotamia around BCE 7500. These cities, such as Eridu, often considered the world's first city in southern Mesopotamia (near the current city of Basra in Iraq) and Uruk were among the many communities between the Euphrates and Tigris rivers.

Ancient cities, such as Faiyum, built around BCE 4000 in Egypt, also formed along the Nile River. In addition, further east, more than 1,000 former urban settlements have been discovered along the Indus River Valley, including the cities of Mohenjo-daro and Harappa in modern-day Pakistan which date back to around BCE 2600. These were among the first cities with diversified economies and societies and were located on trade routes that specialised in gemstones and spanned the whole of central Asia. In terms of its layout, Mohnejo-daro consisted of a clearly planned grid layout with clearly organised buildings and a high level of social organisation focused on a central marketplace. Also, from around BCE 2070, several early cities, collectively referred to as Luoyang, were also formed along the Luo River in what is now Henan Province, China. These examples of some of the world's earliest organised cities included built infrastructure features such as standardised streets and advanced drainage systems.

FURTHER READINGS

Benevolo, L. (1980) *The History of the City*, MIT Press: Cambridge, Mass.
Branch, M. (1985) *Comprehensive City Planning: Introduction & Explanation*, Routledge: London.

Kostof, S. (1991) *The City Shaped: Urban Patterns and Meanings Through History*, Thames & Hudson: London.

Kostof, S. (1992) *The City Assembled: The Elements of Urban Form Through History*, Thames & Hudson: London.

Morris, A.E.J. (1994) *History of Urban Form: Before the Industrial Revolutions*, Third Edition, Longman Scientific & Technical: New York.

Toti, A. & Yang, Z. (eds) (2019) *Grids of Chinese Ancient Cities: Spatial Planning Tools for Achieving Social Aims*, Altralinea Edizoni: Firenze.

ORTHOGONAL CITY

In Europe, many towns and cities created during the Greek and Roman Empires were highly planned using design principles such as order, symmetry, and hierarchy in their layout of streets and spaces together with important buildings. Athens and Rome, for example, are cities that were designed to give a sense of power, control, and grandeur through the formality of their buildings and urban form (Bacon 1975; Morris 1994). Hippodamus of Miletus (480–408 BCE), the ancient Greek scholar, has often been referred to as the 'father of urban planning', and the term Hippodamian plan is still often used to describe a city grid of streets intersecting at right angles to each other (Glaesser 2011). It should be noted, however, that there are precedents for the grid plan in the large empires of the Near East, and also within Greek colonies in the Mediterranean, that precede Hippodamus. Orthogonal urban planning may not have started in Greece; however, they did initiate the organisation of cities into a coherent system of different spaces (Morris 1994). Indeed, reacting to the need to organise sprawling unorganised cities in Ancient Greece, Hippodamus strongly advocated an orthogonal grid as the optimum way to organise urban forms. His plan for Piraeus, the port of Athens, demonstrates a strategic grid layout laid out with wide streets radiating out from a central marketplace or agora, thereby enabling ease of circulation and military efficiency.

During the Roman Empire (BCE 753–CE 476), the Romans also built orthogonal towns or fortified camps, known as castrum, that were either square or rectangular and usually surrounded by a wall with four gates. Two main streets, following the axes created by the four gates, would then divide the town and meet at a central public space or forum. This space, located at the centre of the town, would function as the centre of day-to-day life whilst the rest of the town was then laid out with a regular gridded street pattern that divided the settlements into regular square building plots called insulae (Morris 1994). Initially established as temporary camps, this urban form became the basis for more established towns some of which, such as Marsala in Sicily, still retain evidence of their original layouts. Significantly, the Romans did not just build well-organised cities, they also developed advanced water, drainage, and sewage infrastructure together with high-quality stone-paved streets with separate pedestrian sidewalks to facilitate easy movement and circulation through the settlement.

The ancient orthogonal or grid plan has gone on to be one of the most influential methods of organising and shaping the layout of both new towns and cities, and extensions to existing ones throughout history (Kostoff 1991). One of the biggest influences has been the writings of Vitruvius in his treatise *De architectura* published as *Ten Books on Architecture* and thought to be written between 30 and 20 BCE. Regarded as the first book on architectural theory, it articulates the virtues and fundamental principles of the architecture and urbanism of Ancient Greece and Ancient Rome. The use of the orthogonal grid was also common across China, with cities such as the former capital Chang'an (582 CE) being laid out in an ordered system to demonstrate political symbolism and power.

The Spanish colonisation of the Americas from the early sixteenth century also witnessed the implementation of structured urban planning guidelines for the construction of settlements. The guidelines required that new towns were established with an orthogonal grid of regular plots within a well-ordered layout and streets with a main plaza and a site for a church. The first towns were created in *Hispaniola* (Dominican Republic) from 1502 before rapidly expanding to other areas of the Caribbean and Central America. These early planning guidelines led to the *Ordenanzas de Descubrimiento, Nueva Población y Pacificación de las Indias*, known as The Law of the Indies, being issued in 1573 by King Philip II of Spain. This established strict guidelines for city layouts including regular grid patterns and central public plazas whilst facilitating efficient administration and reinforcing colonial control (Fowler 2022). The requirements included having a detailed master plan for a new settlement in a healthy location with spaces for gatherings and social activities. It also established the importance of having a defined urban fabric with a varied mix of land uses, equality in terms of land distribution, and beautiful architecture.

Contemporary influences of orthogonal or grid planning can be seen in most US cities, including Philadelphia and Chicago, as well as endless others around the world, such as Brasília, Mexico City, Adelaide, Edinburgh New Town, and Milton Keynes (Kostof 1991). Two of the most recognisable contemporary grid patterns are New York and Barcelona. The Manhattan grid, laid out to the Commissioners Plan (1811), was a response to uncontrolled development and health epidemics resulting from cramped irregular streets, together with a political desire for order and convenience. The plan for Barcelona's Eixample district (1860), designed by Ildefons Cerdà (1815–1876), consists of 520 octagonal blocks and straight wide streets enabling ease of movement, sunlight, and ventilation (Hall 1997).

FURTHER READINGS

Kostof, S. (1991) *The City Shaped: Urban Patterns and Meanings Through History*, Thames & Hudson: London.

Kostof, S. (1992) *The City Assembled: The Elements of Urban Form Through History*, Thames & Hudson: London.

Lerup, L. (1977) *Building the Unfinished: Architecture and Human Action*, Sage Publications: Beverly Hills.

Morris, A.E.J. (1994) *History of Urban Form: Before the Industrial Revolutions*, Third Edition, Longman Scientific & Technical: New York.

Owens, E.J. (1992) *The City in the Greek and Roman World*, Routledge: London.

THE VERNACULAR CITY

The vernacular city is an unplanned city, often developed in an informal manner to meet the needs and desires of the local people. Vernacular settlements were predominantly built during the medieval era, often called the Middle Ages, which began around 476 A.D. and lasted for nearly 1,000 years, ending between 1400 and 1450. During this period, in contrast to the grand formality of ancient civilisations, the organisation and layout of settlements were typically unplanned and were dominated by religion and the feudal system. During this time, most people lived in villages and the few towns that existed were typically small and crowded and tended to grow organically as political and economic centres. In Europe and Islamic countries, this resulted in settlements being designed around religious and military buildings together with marketplaces. Vernacular urbanism is typified by patterns of meandering streets and open spaces, and they can be seen as a complex organism in terms of their urban morphology with an informal network of streets and spaces that provided privacy and safety (Kostof 1991). Indeed, vernacular or medieval settlements usually contain unique characteristics that reflect the local contexts of the location and its surroundings such as lifestyles, cultures, geographical features, and climatic conditions. These cities, which were produced without a 'plan' and designed based on rational principles, such as European and Islamic settlements, utilised traditional construction methods and local materials. The buildings also displayed architectural styles and details that were a unique response to the specific attributes of their location. The design of these cities and their architecture is often considered as symbols of the cultural identity of a particular region and reflective of the specific needs and values of the region and its culture.

Vernacular towns and cities contain structures and urban spatial characteristics that are typically created by local people with respect to their own particular environment and socio-cultural beliefs (Mumford 1961). Such settlements are typically created incrementally with relatively small-scale interventions. They also display a number of characteristics such as: the participation of land and building owners in the design and construction process; buildings and spaces that respond to particular local social, cultural, and economic needs and lifestyles; use of locally sourced and affordable construction materials; use of traditional local or regional architectural styles; spatial and building forms that respond to the local climatic and environmental conditions; and circulation layouts that respond to local conditions such as land ownership, established pathways, geographical, and topographical features. Vernacular cities are therefore unique to different places and cultures around the world and are expressions of the identity of a region or place.

In European countries, vernacular towns and cities were often established around a castle or monastery, along a riverbank, or they followed the contour of a hillside. Some others flourished on the sites of early Roman and Saxon settlements, for example, in the UK at London, Winchester, and Canterbury, often having stone walls that were reused and improved. Indeed, almost all medieval towns had perimeter protection with walls, moats, or ditches and gatehouses that controlled entry to the town. Most streets were dirty, narrow, and unpaved whilst being dark, steep, and laid out in an irregular pattern that was difficult to navigate. The streets were also very unhealthy places with no drainage and raw sewage running through them, and only those streets leading directly to a market square might be paved with cobblestones. The only open public spaces, which would serve as gathering spaces, would be a marketplace and those in front of major public buildings such as churches (Kostof 1991). The houses in the town were tightly packed together with the upper floors overhanging the streets. In regions where the houses were of wooden construction, the dense layout often made them susceptible to fire. Many of Europe's surviving medieval towns, such as Carcassonne in France, Bruges in Belgium, and San Gimignano in Italy, are now popular destinations for tourists who value their picturesque townscapes.

In hotter climates, such as the Middle East and Africa, vernacular towns and cities respond to the local climate and environmental conditions and have smaller open spaces and narrower streets to protect inhabitants from the sun and heat. As a result, much of the commercial trading occurs in sheltered environments such as bazaars or covered markets. In addition, the houses tend to be closed to the street and have internal open courtyards that provide privacy and also improve the environmental comfort of the dwellings.

FURTHER READINGS

Kostof, S. (1991) *The City Shaped: Urban Patterns and Meanings Through History*, Thames & Hudson: London.

Mumford, L. (1961) *The City in History: Its Origins, Its Transformations, and Its Prospects*, Harcourt, Brace & World, Inc.: New York.

THE RENAISSANCE AND BAROQUE CITY

The Renaissance period, in the fifteenth and sixteenth centuries, saw a rediscovery of classical design principles resulting in the planning of cities and public spaces that reflected the ideals of humanism. This rebirth of ancient forms led to the grid plan layout being further developed and celebrating the Roman ideal of city planning. Ideal cities were conceived during the Italian Renaissance, when the design of towns prioritised rational urban forms that focusing on human values, urban capacities, and the recursive waves of cultural and artistic revolutions that influenced large-scale planning schemes. The ideal renaissance town with a symmetrical layout centred around a central public space that contained civil buildings (Morris 1994). The fundamental idea of the ideal city was based on a theoretical approach and tended to be planned to ignore geographical and topographical constraints such as rivers, valleys, and mountains. As a result, they were often concepts rather than realised settlements.

Significantly, the ideal city was conceived by Vitruvius during the first century BC in his work *De architectura* but republished during this period by Leon Battista Alberti (1404–1472) in his ten books of treatises on modern architecture called *De re aedifcicatoria* or *On The Art of Building* (1485). He advocated the construction of entire settlements following a perfectly symmetrical order with carefully prescribed rules for the direction and layout of streets, bridges, and town gates. One of the first notable examples of the ideal city includes Filarete's unbuilt proposal for Sforzinda in his *Trattato dell'architettura* written between 1460 and 1464. Planned as an octagonal form with a radial network of streets and a main central square containing a cathedral, castle, and markets. One of the best-realised examples of the Renaissance ideal city is Palmanova in northeast Italy (Hall 1997; Kostof 1991). Constructed as a fortified settlement by the Venetian Republic in 1593, to the design of military architect Friulian Giulio Savorgnan, the layout consists of a symmetrical nine-point star pattern. Designed with a rigid geometric concentric layout, it includes a central square, three nine-sided ring roads, and three main streets radiating out towards the town's entrance gates.

In terms of specific public spaces, the Renaissance period also saw many designs inspired by classical architecture. Michelangelo's reworking of an older public space and surrounding buildings to create the Piazza del Campidoglio (1536–1655) in Rome represents an excellent example of Renaissance design. On Rome's Capitoline

Hill, an imposing wide staircase leads up to a large square with a geometric paving pattern and an equestrian statue of Marcus Aurelius at its centre. Around the piazza, three medieval buildings were given new classical facades to celebrate the city's importance and grandeur whilst representing the Roman ideal of imposing physical order on chaos (Mumford 1961). During the subsequent Baroque period (1600–1750), urban projects surpassed those of the Renaissance in terms of scale and grandeur with its exaggerated emphasis on movement and drama. The redesign of the large square in front of St. Peter's Basilica in Rome by Gianlorenzo Bernini (1598–1680) exemplifies this style. Completed in 1667, the huge oval square centred on an obelisk has three centres surrounded by two imposing semi-circular colonnades that embrace the Basilica and open up to the city at the opposite side.

During the Baroque period, many cities expanded and removed medieval city walls creating dynamic baroque projects across Europe. An overriding aim was to create cities as works of art that symbolically represented the Catholic Church's authority with churches becoming the focal points and landmarks of the city. In doing so, long streets and grand unified open spaces were formed and integrated with elaborate classical-inspired architecture and sculpture. Baroque architects and artists expanded upon classical Renaissance traditions and whilst still emphasising symmetry and harmony introduced heightened dynamism, grandeur, and power and to heighten emotional and sensual experience. To achieve this, the arrangement of buildings and streets would exploit perspective and visual illusions to enhance the sense of depth and also utilise focal points, such as sculptures and obelisks, and key buildings to terminate vistas (Bacon 1975).

Two projects that exemplify the Baroque approach to urban spaces in Rome are the Piazza Navona and the Piazza del Popolo. Piazza Navona was built on the perimeter of an ancient Roman stadium resulting in its long and narrow proportions. Surrounded by examples of Baroque architecture the piazza also contains significant examples of Baroque sculpture including works by Bernini, and Borromini. The Piazza del Popolo was laid out in 1538 as a grand entrance to Rome's northern gateway to the city and has subsequently been remodelled many times. The most recent by Guiseppe Valadier (1726–1785), between 1811 and 1822, created its elliptical plan with a large obelisk at its centre and neo-classical architecture. Twin churches by Carlo Rainaldi (1611–1691), that precede Valadier's design, frame the southern approach to the square which is conceived as a forecourt for arrival in the city.

The ideal city was seen as a utopia, and many utopian ideals for society and the city emerged during the Renaissance period. Indeed, utopia was considered to be a place where there was perfection in the whole of society. This concept originated in Sir Thomas More's (1478–1535) book *Utopia* which was published in 1516. In the book, More describes an imaginary world in which people living in a utopian society share a communal culture and way of life. A significant figure in the English Renaissance, his aim was to show how European society could be improved and corruption, poverty, and scandal could be replaced by a self-governed, democratic, and equal ideal society. In *Utopia*, More also described the design and layout of a city as a geometric form surrounded by a wall similar to all Renaissance ideal city plans. Significantly, many of the concepts embodied in utopian society and the ideal city have continued to inspire designers of new towns and cities through subsequent centuries.

FURTHER READINGS

Bacon, E.N. (1975) *Design of the Cities*, Revised Edition, Thames & Hudson: London.

Hall, T. (1997) *Planning Europe's Capital Cities: Aspects of Nineteenth Century Urban Developments*, E & FN Spon: London.

Morris, A.E.J. (1994) *History of Urban Form: Before the Industrial Revolutions*, Third Edition, Longman Scientific & Technical: New York.

THE BEAUTIFUL CITY

The urbanisation resulting from the Industrial Revolution totally transformed cities. In the UK, for example, in 1801 around twenty per cent of the population lived in towns and cities; however, by 1851 over half of the population was considered urbanised increasing to nearly seventy-five per cent by 1901 (Wrigley 1985). This period of rapid economic development led to an unprecedented expansion of cities in order to meet the industrial needs for manufacturing, transportation, and housing for large workforces. The impact of this often-unplanned development and growth led to the destruction of traditional communities, slum housing, pollution, disease, and poor sanitary conditions becoming the norm in developing industrial boom cities across the UK, Europe, and North America.

One of the most significant urban planning projects of the nineteenth century was the deconstruction and reorganisation of Paris. At the time, Paris was a rapidly expanding with chaotic and overcrowded medieval city typified by poor living conditions, disease, and poverty. Ruler of the Second Empire (1852–1870), Napoleon III, had been impressed by the wide streets, parks, and infrastructure of London and wanted to turn Paris into a new imperial city that celebrated the glory of the French empire. As a result, in 1853, he commissioned Baron Georges-Eugene Haussmann (1809–1891) to modernise the city and improve the quality of life in Paris by creating wide streets, public parks, and reservoirs (Bacon 1975; Hall 1997). The most ambitious aspect of Haussmann's plans was the reorganisation and planning of the city with ancient meandering streets replaced by wide straight boulevards flanked with imposing buildings and bridges over the River Seine that were not only grand statements but also enabled the free movement of troops (Kostof 1991). The resultant grand radiating boulevards created vistas and promenades through the city that were associated with the clarity and rationale of the French Enlightenment (Mumford 1961). Haussman's 20-year plan, which involved demolishing around 20,000 historic buildings and constructing more than 34,000 new structures transformed the city. He cut bold new arteries through the cramped and chaotic labyrinth of slum streets and cleared space for significant new public buildings such as the Palais Garnier opera house, Les Halles market hall, the Gare du Nord and Gare de L'Est, and connected these with long, wide, and straight boulevards. He also created a new sewage network together with reservoirs and aqueducts bringing clean drinking water to the city (Hall 1997). This brought modernity and status to the city and gave it a sense of grandeur that is still evident today. However, many remain divided over whether the transformation of Paris into the City of Light is the masterpiece of a great master planner or the work of an imperialist megalomaniac.

In the United States, as cities expanded and became increasingly industrialised in the late nineteenth century, new approaches to city planning were sought in order to improve living conditions in urban slums and create spaces of beauty and grandeur. One significant approach saw the emergence of the City Beautiful movement, which, was concentrated in the United States, but was inextricably linked with other philosophies around the world such as the Garden City movement in the UK. The City Beautiful movement was not just an aesthetic style and incorporated ideals that included order, harmony, and structure that reflected the classical architecture, geometric layouts, broad boulevards, and imposing monuments that characterised the grandeur of many European cities (Morris 1994). Indeed, proponents of the City Beautiful philosophy considered it as a way to improve public health and social well-being through good design and orderly city planning where beautiful public spaces would lead to social harmony and civic pride.

One of the biggest influences of the City Beautiful approach to urban planning in the United States can be seen in the plan (1791) by Pierre Charles L'Enfant (1754–1825) for Washington, D.C. In the plan, L'Enfant designed a new capital with wide avenues, public squares, and monumental buildings with the National Mall as its focal point. Capitol Hill was designed to be the centre of the city with diagonal avenues radiating out and cutting across a grid plan of streets. These wide boulevards facilitated transportation and also presented grand views of key buildings and public squares. L'Enfant's plan was not fully implemented and was later updated by the McMillan Plan (1902) which was developed by the McMillan Commission, which included Daniel Burnham, landscape architect Frederick Law Olmsted Jr., and architect Charles McKim. The plan revived and expanded upon many of L'Enfant's original concepts with the aim of beautifying the nation's capital city.

Designs that embodied the spirit of the City Beautiful were first revealed at the 1893 World's Columbian Exposition in Chicago with a project by Daniel Burnham (1846–1912). His 'White City' was a sprawling prototype for a new American city with neoclassical structures to create a vision of balance, symmetry, progress, and social harmony. Burnham believed that city planning should be a tool for social engineering that could inspire good citizenship. The City Beautiful was inspired by Haussmann's Paris and further developed in plans to beautify and add monumental grandeur to cities such as Cleveland (Group Plan, 1905) and Chicago (Plan of Chicago, 1909) to plans by Burnham. The culmination of the City Beautiful Movement came with Bennett and Day's plan for Detroit's Center of Arts and Letters (1913) which created a new cultural centre with public buildings for the city. This included the development of many Italian Renaissance and Beaux-Arts-style buildings, such as the Public Library, Institute of Arts, and Orchestra Hall, by eminent American architects. Subsequently, critics such as Jane Jacobs (1961) have described the City Beautiful approach as overly cosmetic, whilst others have suggested that it proved expensive and its focus on civic architecture neglected the lives of normal citizens and areas beyond the civic core.

In the UK, over half of the population lived in towns in the middle of the nineteenth century, and by 1900, it had risen to over three quarters. As a result, towns and cities suffered from social and environmental problems, such as slum housing, poor sanitation, pollution, and congestion, of unprecedented scale and urban planning efforts of the time focused upon the terrible living conditions of most urban residents (Ward 2011). The Garden City Movement, with a much greater emphasis on people and communities, emerged out of these issues following the publication of Ebenezer Howard (1850–1928)'s publication *To-Morrow: A Peaceful Path to Real Reform* (1898). In this, Howard proposed a garden city as a hybrid of the best of both town and country to deliver a *"joyous union"* from which *"will spring a new hope, a new life, a new civilisation."* The book was subsequently reprinted in 1902 as *Garden Cities of Tomorrow*. Howard's visionary ideas were inspired by Arts and Crafts values and the various nineteenth-century model village projects by industrial philanthropists, such as New Lanark (1800) near Glasgow, Saltaire (1851) near Bradford, Port Sunlight (1888) on the Wirral, and Bourneville in Birmingham (1893).

In his Garden City proposal, Howard illustrated his ideas with a circular urban plan that offered the societal and economic benefits of the city while surrounding inhabitants with the natural beauty of the countryside. He advocated self-contained communities that were surrounded by greenbelts but connected to each other by public transport, thereby avoiding reliance on traditional urban centres. There would be a series of self-sufficient garden cities with a population of 32,000 organised in healthy and sociable communities of walkable neighbourhoods with high-quality affordable housing, local employment opportunities, public parks, good infrastructure, and easy access to natural landscapes. Whilst being domestic and human in scale, the proposal also advocated a central park with public buildings, retail, and commerce with crescents and terraces around a broad circular 'Grand Avenue'. The first realisation of the Garden City concept was at Letchworth in the UK when construction began in 1903. Designed by Barry Parker (1867–1947) and Raymond Unwin (1863–1940) the master plan for the new community that embodied many of Howard's ideas whilst adapting his circular diagram to local topography and landscape (Kostof 1991). Parker and Unwin subsequently designed Hampstead Garden Suburb (1907) in North London, which was not strictly a garden city but strongly influenced by the concept, and Welwyn Garden City (1920) which was founded by Howard and designed by Louis de Soissons (1890–1962).

With its widespread appeal, Howard's Garden City ideas spread globally and inspired the design of settlements across the world, including in America, Canada, Asia, South America, Australia, New Zealand, South Africa, and Europe. Like most radical and visionary urban planning ideas, the Garden City also had its critics, with some suggesting that it was anti-industry during a time of industrial growth and also that its approach was damaging to nature, given a focus on designed green spaces and low-density neighbourhoods. Nevertheless, the Garden City concept has remained influential to urban planning and more recently urban design theory and practice, from the UK's New Town Movement through to the US-originated New Urbanism movement that promotes environmentally friendly places and walkable neighbourhoods (Rudlin & Falk 2010).

FURTHER READINGS

Hall, T. (1997) *Planning Europe's Capital Cities: Aspects of Nineteenth Century Urban Developments*, E & FN Spon: London.

Howard, E. (1898) *To-Morrow: A Peaceful Path to Real Reform*, Swan Sonnenschein: London. Republished as *Garden Cities of To-morrow* in 1902, Faber and Faber: London.

Kostof, S. (1991) *The City Shaped: Urban Patterns and Meanings Through History*, Thames & Hudson: London.

Morris, A.E.J. (1994) *History of Urban Form: Before the Industrial Revolutions*, Third Edition, Longman Scientific & Technical: New York.

Ward, S. (2011) *The Garden City: Past, Present and Future*, Spon Press: London.

NEIGHBOURHOOD UNIT TO NEIGHBOURHOOD PLANNING

The neighbourhood unit concept can be traced back to the nineteenth century when many scholars were concerned about the quality of urban life. It was, however, the publication of Clarence A. Perry (1872–1944)'s *The Neighbourhood Unit* in 1929 that promoted the approach as a comprehensive planning tool. Concerned with the poor living conditions in rapidly industrialising New York in the early 1900s, Perry advocated self-containing residential areas with a community-focused lifestyle, away from the busy city core. These interconnected neighbourhoods were proposed as a collection of small-scale communities with residential units located close together sharing activities, amenities, and facilities, which he believed would create a sense of belonging.

Perry later published his book *Neighborhood Planning* (1953) which further stressed the significance of the physical environment in creating a sense of community. He argued that quality of life is dependent on the quality and design of housing, streets, paths, and landscape. Proposing a community focus to design, he argued that neighbourhood planning should be based on the needs of residents, and in doing so, this will facilitate social interaction and cohesion. Perry's ideal neighbourhood would consist of about 5,000–9,000 residents with good public transport and contain schools, religious buildings, and recreational areas.

Perry's neighbourhood unit theory was subsequently widely adopted for the basic planning of cities around people-focused communities. The concept did, however, draw some criticism, particularly for being too restrictive in terms of land use, suggesting that communities with a greater mix of land uses were preferable. Others have argued that the neighbourhood unit did not take into account the diversity of cultures and lifestyles across different regions and therefore not applicable across all parts of the world. Despite some suggesting that the neighbourhood unit can help reduce the level of poverty and inequality in the society, the concept has also been criticised as a tool to enable social segregation into homogeneous groups. Despite these concerns, neighbourhood planning was rediscovered in the 1960s, with a move by urban planners to engage and plan with communities to fix existing neighbourhoods as opposed to the trend for large-scale urban renewal and redevelopment projects.

FURTHER READINGS

Banerjee, T. & Baer, W.C. (1984). *Beyond the Neighborhood Unit*. Plenum Press: New York.

Lawhon, L.L. (2014) Neighborhood Unit. In: Michalos, A.C. (ed.) *Encyclopedia of Quality of Life and Well-Being Research*, pp. 4335–4337, Springer: Dordrecht. https://doi.org/10.1007/978-94-007-0753-5

Mumford, L. (1954) The Neighborhood and the Neighborhood Unit, *The Town Planning Review*, 24(4), 256–270. www.jstor.org/stable/i40003315

THE MODERNIST CITY

The on-going issues in cities, such as overcrowding, poor living conditions, inadequate transport infrastructure, and pollution, continued into the early twentieth century. In response, a group of influential European architects, led by Le Corbusier (1987–1965), formed *The Congres Internationaux d'Architecture Moderne* (CIAM) in 1928. The aim of the organisation was to advocate the principles of modern design into the realms of urban planning and city design, in order to create better and more functional places (Gold 1997; Mumford 2018). CIAM's fourth conference saw the launch of their manifesto, the *Athens Charter* (1933), which proposed that the social problems of cities be resolved through strict functional segregation planning. The charter was based on Le Corbusier's unrealised urban master plan first presented as *Ville Contemporaine* in 1922 and later published as *Ville Radieuse* (1935).

Celebrating the modernist ideals of progress, the *Radiant City* was designed to be constructed on the remains of demolished vernacular European cities. Designed around fast and efficient transportation, the new city planned to a Cartesian grid and designed as a functional 'living machine' would contain identical high-density skyscrapers spread across a vast green landscape. Le Corbusier's proposed city was divided into functionally segregated into commercial, business, entertainment, and residential districts. Like most other late-nineteenth and early-twentieth century solutions to the problems of urbanisation he argued that his proposal would provide residents with a better lifestyle and contribute to the creation of a better society. Le Corbusier's radical ideas were further developed in his proposals for projects for Paris, Antwerp, Moscow, Algiers, and Morocco. Eventually, in the Indian city of Chandigarh, in which Corbusier was involved from 1950, he was able to build a modernist city embodying his ideals with his design for the new capital of the Punjab. The planned capital city of Brasilia, planned by Lúcio Costa (1902–1998) in the 1950s, is one of the best examples of CIAM's ideals on city design and planning. Designed as a modern utopia, the city was planned around a repetitive series of mono-functional superblocks. These are organised around a monumental east–west axis of political and administrative functions, and a north–south residential axis that divided the city. The simple aesthetic and oversized scale of its modernist architecture together with the broad vistas of the roads and landscape enhanced the feeling of grandeur and monumentality (Kostof 1991).

Around the same time in the United States, American architect Frank Lloyd Wright (1867–1959) outlined his proposal for a new dispersed and automobile-based city in his book *The Disappearing City* (1932). A few years later, in 1935, he presented a more detailed proposal called Broadacre City which would be a decentralised city based around the free-flowing movement of the car. With an aim of achieving a democratic, integrated, and organic settlement, Wright proposed that four square miles of countryside could become a city with 1,400 families at an optimum density of 2.5 people per acre. All families would have a minimum of one acre of land and the city would be divided by landscaped highways. He also suggested that all of the homes would be to different designs and that the buildings would be constructed of steel and glass to enable access to sunlight, air, and views. Wright's radical ideas that planning and architecture could solve all of society's problems were impractical and also failed to foresee the imminent population explosion. As a result,

Broadacre City was never realised; however, many of its characteristics are typified in the low-density urban sprawl witnessed in the United States in the late twentieth century.

Despite the radical, strict, and nearly totalitarian standardisation, order, and symmetry of CIAM's proposed principles, they had a significant influence on modern urban planning and high-density housing typologies particularly after World War II, with many of Europe's major cities requiring significant reconstruction and new housing (Gold 1997). Despite significant criticism in the late twentieth century, there have been many celebrated examples of successful modernist urban projects at an architectural scale. For example, the Allt-Erlaa (1975–1986) public housing complex in Vienna designed by Harry Glück (1925–2016) demonstrates how the modernist approach can be successful if designed well and in the correct context. Similarly, although not an affordable housing project, The Barbican Estate (1965–1975) designed by Chamberlin, Powell, and Bon in the City of London demonstrates how brutalist architectural projects can succeed in the right location and when they integrate a diverse range of land uses and activities, are well-connected, and well-maintained.

The modernist approach, to city planning at least, was however generally regarded as a failure with CIAM being disbanded in 1959. It is argued that modernist city plans failed to show an understanding of communities and neighbourhoods, and that planned large open spaces showed no consideration of how people might use them. Indeed, Allan Jacobs (1993: 111) identifies that: "As well-intentioned and socially responsive as those manifestos were, there results, abundantly visible by the 1960s, rarely encourage or celebrate public life. They seem more consistent with separation and introspection…" Significantly, the 1972 demolition of the Pruitt-Igoe estate (1951–1955) in St Louis, Missouri, which had been designed to CIAM's functional city ideals, was claimed by architecture critic Charles Jencks to be the death of the modern architecture and city design movement.

FURTHER READINGS

Gold, J. (1997) *The Experience of Modernism: Modern Architects and the Future City*, Routledge: London.

Lang, J. (2021) *The Routledge Companion to Twentieth and Early Twenty-First Century Urban Design: A History of Shifting Manifestoes, Paradigms, Generic Solutions, and Specific Designs*, Routledge: New York.

Mumford, E. (2018) *Designing the Modern City: Urbanism Since 1850*, Yale University Press: New Haven.

NEW TOWNS

In the United Kingdom, building upon the legacy of the Garden City Movement, a programme of creating new settlements under The New Towns Act 1946 led to the delivery of 32 New Towns across the UK (Hardy 2012). In response to overcrowding and extensive bomb damage experienced in cities during WWII, the most ambitious town building programme ever undertaken in the UK was completed in three phases between 1946 and 1973. Later developments included the expansion of existing towns to accommodate overspill population from densely populated urban areas. Early settlements focused on addressing housing shortages, in the first phase these included Stevenage, Hemel Hempstead, and Harlow, whilst the second phase included Skelmersdale and Runcorn. The third phase generally focused on enabling further population growth through the creation of much larger new towns such as Milton Keynes, together with the expansion of existing towns such as Northampton and Peterborough.

The design-led ideology behind UK's New Towns included innovations such as dedicated public transport routes to facilitate efficient and affordable circulation, solutions to promote walkability like underpasses, pedestrianised town centres with car-free shopping streets, as well as the UK's first shopping malls. Green spaces were also key following the Garden City experiment of the 1900s. Indeed, the low-density housing had grass verges and front gardens together with public green spaces and parks separating residential neighbourhoods referenced the garden cities of the early twentieth century. Similarly, a green belt for agricultural and leisure activities which also segregated industrial zones, to protect from noise and pollution, was planned for each town. The residential areas, which included considerable social housing, were organised around amenities like schools, medical centres, leisure complexes, and community centres (Alexander 2009).

Often described as soulless and boring, the New Towns were a solution to the housing crisis after WWII. However, with many built surrounding London, their location meant that affordability, rail connections, and proximity to London resulted in many of the New Towns became commuter towns which conflicted with the key aim of being self-contained economic and social units. Also, despite the creation of pedestrian-friendly environments, the growth in car ownership together with the low density of development led to high levels of car dependency. The New Towns did, however, inspire town planners and architects from around the world with new towns subsequently being developed in many European countries such as Finland, France, Germany, Italy, Netherlands, and Sweden (Lock & Ellis 2020). The movement also had a wider global influence upon new settlements created in Hong Kong, Japan, South Korea, and the United States, amongst others.

FURTHER READINGS

Alexander, A. (2009) *Britain's New Towns: Garden Cities to Sustainable Communities*, Routledge: London.

Lock, K. & Ellis, H. (2020) *New Towns: The Rise, Fall and Rebirth*, RIBA Publishing: London.

THE COMPACT CITY

The compact city refers to an urban model typified by medium-to-high residential density and mixed-use neighbourhoods that promote walking, cycling, and public transportation. The term was actually first used in *Compact City: A Plan for a Liveable Urban Environment* (1973) by two American mathematicians, George Dantzig and Thomas Saaty, in which they advocated a utopian vision that would enable more efficient use of resources. In terms of its influence upon urban planning, the concept builds upon many of the ideas expressed by Jane Jacobs in her book *The Death and Life of Great American Cities* (1961). Recognising the qualities of existing dense urban communities and seeing the negative impacts of both urban renewal projects and low-density solution, Jacobs strongly argued for mixed uses, small walkable blocks, building of different ages and types, and a dense concentration of homes (see also Chapter 3).

The concept of the compact city is considered as a more sustainable urban model that focuses upon providing everything people need to live in one community. It is therefore associated with a more densified urban pattern where living, working, shopping, socialising, and leisure opportunities are co-located to facilitate the convenient and safe movement of pedestrians, cyclists, and public transport users (Burton et al. 1996, 2000). Implemented successfully, the concept would therefore improve accessibility and make communities more inclusive and equitable whilst reducing fossil fuel consumption, emissions, pollution, and traffic density. A popular concept with European urban planners, the cities of Amsterdam and Copenhagen are known examples of such a model, whilst it also inspired new developments such as Poundbury in the UK. The term has also influenced the development of the smart city concept, which is similar, but recognises the necessity of urban growth. The compact city idea also has clear synergies with the subsequent ideas advocated by the New Urbanism movement and the 15-minute city model.

FURTHER READINGS

Burton, E., Jenks, M. & Williams, K. (eds) (1996) *The Compact City: A Sustainable Urban Form?*, London: Routledge.

OECD (2012) *Compact City Policies: A Comparative Assessment,* OECD Green Growth Studies, The Organisation for Economic Co-operation and Development, OECD Publishing. www.oecd.org/content/dam/oecd/en/publications/reports/2012/05/compact-city-policies_g1g191f1/9789264167865-en.pdf

NEW URBANISM

In the United States, in the mid-to-late 1980s, many built environment professionals were frustrated with post-WWII urban development patterns that was focused on land-use zoning, low-density, and urban sprawl rather than traditional urban models (Duany et al. 2000). They also observed that urban renewal projects were destroying historic neighbourhoods and longstanding communities (CNU 2024). As a result, a group of architects, urban planners, urban designers, developers, and engineers formed the as a non-profit organisation in 1993. The Congress was established as *"...a movement for reinvestment in design, community, and place"* (CNU 2024). A key aim of the Congress was to facilitate the creation of well-designed neighbourhoods, towns, and cities that foster healthy and thriving communities. The promotion of social, economic, and environmental opportunities was advocated through the creation of liveable, sustainable, and walkable communities, as a result of managed growth and traffic reduction (CNU & Talen 2013).

In physical terms, the New Urbanism approach is based on human-scale and historical principles of how cities and towns have been constructed. As such, it strongly advocates the creation of walkability with small urban blocks, pedestrian-friendly streets that prioritised walking, cycling, and public transportation, mixed-use development through the co-location of a diverse range of homes, shops, and facilities, and well-connected and accessible public spaces that encourage daily interaction and public life (Katz 1994). The Congress for New Urbanism (2024) emphasises that these *"principles can be applied to new development, urban infill and revitalization, and preservation [and] ...all scales of development in the full range of places including rural Main Streets, booming suburban areas, urban neighborhoods, dense city centers, and even entire regions."*

Since the late twentieth century, there have been many examples of New Urbanist projects delivered around the world. For example, the master-planned town of Seaside (1981–1985) in Florida, designed by Andres Duany and Elizabeth Plater-Zyberk, was a pioneer community in terms of adopting New Urbanist principles. Developed as an 80-acre resort community on the shores of the Gulf of Mexico, Seaside's master plan aimed to reflect the character of an old Southern town. As such, it has a network of small-scale walkable streets that intersect to create small urban blocks and provide a comfortable and safe walking and cycling environment. The streets are also designed to be mixed-use with homes, shops, and restaurants all within walking distance. This and the fact that the town's streets are well-connected to neighbouring settlements ensures a vibrant atmosphere throughout the day. Significantly, a simple form-based code was implemented to inform the design and architectural features of all of the town's buildings. This included, for example, that all houses had to have front porches and picket fences. Like most pioneering urban ideas, Seaside has drawn criticism for being too expensive, too exclusive, too controlled, too historicist, etc.

Despite being a vacation community rather than a realistic urban settlement, Seaside has, however, served as a model and inspiration for subsequent towns across the United States and around the world. Indeed, new settlements and urban-scale developments in many countries, including Argentina, Canada, China, Colombia, Costa Rica, France, Guatemala, Iran, Mexico, Panama, and Sweden, have been heavily influenced by the principles of New Urbanism. In the UK, for example, Poundbury (1993–2025) was developed as an experimental urban extension to the market town of Dorchester in Dorset. Influenced by an approach similar to New Urbanism, the town is built according to the principles of its patron King Charles III, who is known for strong views against post-war trends in town planning. The master plan for an urban village with a population of 6,000 population at 15–20 dwellings per acre was designed by urban theorist and planner Léon Krier (1946–). In creating a walkable and sustainable community, the organisation of the buildings predetermined the road layout with winding streets connected by pedestrian alleyways. Designed as a high-density mixed-use urban pattern that integrates private and social housing with shops, and businesses, the town is designed around people rather than motor vehicles. Allowing convenient local access to amenities, the design of Poundbury was an early example of a 15-minute neighbourhood that pre-dates the 15-minute city concept. It also followed many of the ideas for a compact city as recommended in the UK Urban Task Force's *Towards an Urban Renaissance* (Urban Task Force 1999). An architectural code for the town that favoured tradition was created by Andres Duany and resulted in the centre being built in a classical style and outer residential areas in a vernacular style. Critics of the development have focused upon the reliance on historic aesthetics and the creation of a 'sanitised' settlement; however, Poundbury continues to appeal to many and be a sought-after residential location.

FURTHER READINGS

CNU & Talen, E. (ed) (2013) *Charter of the New Urbanism*, 2nd edition, McGraw Hill: New York.

Congress for New Urbanism. www.cnu.org

Duany, A., Plater-Zyberk, E. & Speck, J. (2000) *Suburban Nation: The Rise of Sprawl and the Decline of the American*, North Point Press: New York.

Haas, T. (ed.) (2008) *New Urbanism and Beyond: Designing Cities for the Future*, Rizzoli: New York.

Katz, P. (1994) *The New Urbanism: Toward an Architecture of Community*, McGraw-Hill: New York.

THE 15-MINUTE CITY

The concept of the 15-minute city builds upon earlier urban planning models such as Ebenezer Howard's *Garden Cities of To-morrow* (1898) and Clarence Perry's *The Neighborhood Unit* (1929). It's also comparable to the compact city concept, an urban planning model first mentioned in the 1970s but often attributed to the ideas of Jane Jacobs in *The Death and Life of Great American Cities* (1961). Although the idea of the mixed-use walkable neighbourhood has been promoted by urban designers for many years, the 15-minute city concept has gained significant traction after it was first proposed at the COP21 conference in Paris (December 2015) by Colombian-French professor Carlos Moreno (1959-). Subsequently, Paris mayor Anne Hidalgo proposed the implementation of the concept in her 2020 re-election campaign and the popularity of the concept grew in the wake of the COVID-19 pandemic and its impact upon urban communities. Significantly, it has rapidly influenced urban planning policy and practice as other cities around the world have begun to embrace the idea in the search of more sustainable cities.

The 15-minute city is an urban planning model where residents have easy access to daily needs within a 15-minute walk or cycle ride. Achieving this can lead to more accessible, inclusive, and sustainable cities with reduced car use, can be created that enable people to live well locally in human-centric communities that facilitate quality interactions. Moreno (2021, 2024) argues that there are six essential urban social functions that all inhabitants should have access to within 15 minutes of their homes: *"Living, working, supplying, caring, learning, and enjoying (i.e. housing, work, food, health, education, and culture and leisure)."* To achieve this involves designing local areas to have daily necessities and services, such as work, shopping, education, healthcare, and leisure within this distance of where they live (Moreno 2024). There should also be affordable, convenient, and regular public transportation from any point in the city. Although equally applicable to new settlements, the 15-Minute City concept is more likely to be applied to the transformation of existing cities. Importantly, its implementation will require a multi-disciplinary approach involving urban planners, urban designers, transportation planners, policymakers, and communities, if the goals of enhanced well-being through inclusive, accessible, diverse, healthy, and sustainable environments are to be achieved. It is also important to note that the notion of 15 minutes should not be seen as a fixed time, and many similar concepts, such as the 20-minute city (Larson 2012), 20-minute neighbourhood, one-mile city (D'Acci 2013), the German stadt der kurzen Wege (city of short distances), the complete community, and the city village, have also been advocated in recent decades. Despite many cities across the world, such as Paris, Melbourne, Barcelona, Bogota, Ottawa, Portland, and Shanghai, having adopted the concept, the 15-minute city has drawn some criticisms such as the potential to favour more affluent neighbourhoods and segregate others by income or race.

FURTHER READINGS

Luscher, D. (n.d.) *The 15-Minute City Project*. www.15minutecity.com

Moreno, C. (2024) *The 15-Minute City: A Solution to Saving Our Time and Our Planet*, Wiley: New York.

Whittle, N. (2021) *The 15-Minute City: Global Change Through Local Living*, Luath Press: Edinburgh.

Whittle, N. (2024) *Shrink the City: The 15-Minute City Urban Experiment and the Cities of the Future*, The Experiment: New York.

SUMMARY

In the twenty-first century, most urban design projects involve engaging with existing urban contexts, and where they are concerned with the design and creation of new settlements, these will still have local or regional relationships with existing towns and cities. It is therefore imperative that urban designers have a thorough understanding of how existing urban forms have evolved throughout their history together with the influences that have informed their creation and design approach. This chapter has therefore focused chronologically upon approaches and theories behind the creation of public spaces and more broadly urban forms to enable a contextual understanding of the application of these various approaches. Although this chapter has been organised into discreet approaches to city and public space design, in reality most settlements have been impacted by multiple periods, approaches, and styles during their evolution. Some towns and cities might display dominant characteristics; however, most will demonstrate a fusion or mix of influences representing different eras of growth and change.

In addition to understanding the historic context of the public realm, greater knowledge of previous and existing approaches to the creation of urban form enables inspiration to be drawn from historical precedents (Krier 2006). Indeed, throughout history, many successful public spaces have drawn influence, at least from a physical perspective, from design solutions from other eras. Prior to the twentieth century, much of the attention of public space design focused upon the physical, aesthetic, and practical dimensions of space. This evolved during the twentieth century and in the twenty-first century has led to a more inclusive and sensitive consideration of the needs and requirements of people and the planet. This chapter has therefore also highlighted how the emphasis on public space creation has become much more human-centric and community-focused to address contemporary challenges like resilience and sustainability.

PLACEMAKING FROM THEORY TO PRACTICE

The concept of place and the genius loci of urban spaces have become central to the practice of contemporary placemaking. Despite placemaking, which can be considered as both a philosophy and a process, not being commonly used as a term until the mid-to-late 1990s, its foundations can be traced to the early twentieth century. Until this time, much of the focus on urban space was concerned with its physical attributes. Indeed, Giambattista Nolli (1701–1756) was one of the first to research the city, when he used a figure-ground diagram to represent complex connections and relationships between public spaces and private buildings in his 1748 Nolli map of Rome (Verstegen & Ceen 2013). Previously, cities tended to be artistically represented using pictorial maps or bird's eye views to emphasise landmarks rather than accurate maps. These were of little use in terms of wayfinding and urban management and the Nolli map demonstrated an understanding of the patterns of the built form and the continuity of public space that provided an important tool for the analysis and design of new projects.

In the early twentieth century, a growing interest in the human and sociological dimension of cities and the public realm began to emerge with luminaries such as Patrick Geddes, through to the significant observations and theories of Jane Jacobs, Gordon Cullen, Kevin Lynch, William H. Whyte, and others. Many of these emerged as a reaction to criticisms of grandiose schemes of the 1920s and 1930s and modernist projects of the 1950s and 1960s.

In addition, there was growing recognition of the importance of understanding how people experience the city, celebrating the vitality of urban neighbourhoods, how communities could be engaged in the planning process, and human-centric design. Continued erosion of the public realm in cities through the 1970s and 1980s through car-dominated, mono-functional planning and zoning led to further efforts to focus urban planning and design towards a more sensitive approach towards the existing urban fabric and to the patterns of behaviour of its inhabitants. As one of the eminent theorists and practitioners, Jan Gehl (2010) discussed how the traditional city had been invaded by vehicles, subsequently abandoned, and was now being reconquered as a place for people once again. Indeed, Gehl, his contemporaries, and organisations such as the Project for Public Spaces have made people-focused concepts such as mixed-use neighbourhoods, walkable streets, and vitality in public places central to the practice of urban design. As a result, urban designers now commonly use the terms place and placemaking to promote urban projects whilst increasingly aiming to facilitate the participation of people and communities in the design process.

This chapter will therefore examine changing approaches and attitudes to place and placemaking. This will articulate how the initial focus on the design of urban spaces was on their physical qualities and organisation within cities. Subsequently, in the early twentieth century, the complexities and importance of public spaces within urban environments began to be recognised with increasing attention on the intrinsic relationships between place and people. As a result, place began to be researched and understood from a range of varying perspectives, including its sociological, ecological, psychological, and behavioural dimensions. This chapter will, therefore, focus on many of the key figures in this evolution of the understanding of place through to contemporary thoughts and approaches to placemaking.

FURTHER READINGS

Koch, R. & Latham, A. (eds) (2017) *Key Thinkers on Cities*, Sage Publications: Los Angeles.

Larice, M. & Macdonald, E. (eds) (2012) *The Urban Design Reader*, Routledge: London.

Project for Public Spaces (2008) *Placemaking Heroes Articles.* www.pps.org/category/placemaking-heroes

Tiesdell, S. & Carmona, M. (eds) (2007) *Urban Design Reader*, Routledge: Oxford.

CAMILLO SITTE

One of the first to articulate theories related to the city and its public spaces was nineteenth-century Austrian architect and town planner, Camillo Sitte (1843–1903). In his seminal book, *Der Städtebaunach seinen künstlerischen Grundsätzen* (1889), translated into English as *The Art of Building Cities: City Building According to Its Artistic Fundamentals* (1945), Sitte presented a critical analysis of streets and squares in many European cities that retained historic centres. Being highly critical of the patterns of industrial urbanism in Europe, he was motivated to undertake the study having witnessed the impact of the Industrial Revolution upon the economic, physical, and social transformation of cities. Sitte was dismayed by how modern city planning focused on solving the city's challenges as a technical problem rather than as an art and considered that this was resulting in symmetrically balanced projects that lacked were the spatial sensitivity of pre-industrial cities. As such, he challenged the trend towards rigid symmetry and the placement of public buildings as objects in space on large plots. He also suggested that modern city planning failed to recognise the proper relationship between public spaces and buildings by creating monotonous urban patterns with only the leftover space being for streets and squares. In contrast, he advocated traditional approaches for the creation of public spaces inspired by traditional city planning traditions. This inspired Sitte to develop, in his book, a set of artistic principles to guide to the design of new urban areas.

In his research, Sitte used detailed sketches and maps to analyse the relationships between public squares and adjacent buildings. Significantly, he identified that that in cities developed during the Middle Ages and the Renaissance, public spaces were vital for public life and that there was a strong interdependency between squares and surrounding public buildings. Sitte also suggested that the architecture had a fundamental role to play in providing physical enclosure to the public realm and likened successful public spaces to enclosed rooms. He also strongly argued that the aesthetic experience of urban spaces should be the leading consideration in the design of cities. The examples presented by Sitte also helped to establish an understanding of the city and the intricate interrelations of its parts. In doing so, he also emphasised the value of irregularity in the urban form and carefully articulated the role and purpose of both irregular and straight streets, together with the value of creating a rich pedestrian-oriented public realm. Sitte also introduced the importance of experiencing the city through sequential movement and argued for the creation of visually stimulating urban environments with public space being well-enclosed and well-defined by its encompassing buildings.

At a time, when cities were facing significant challenges to accommodate growth, with new approaches to building density, land-use patterns, and transport infrastructure, Sitte's observations were both innovative and controversial. Interestingly, despite proposing different solutions, many of Sitte's ideas and observations were similar to those of his contemporary Ebenezer Howard (see Chapter 2) who was proposing his Garden City concept at a similar time. Although the initial reaction to his work was positive, support declined with the emerging dominance of a modernist view of the city. Nevertheless, despite his focus mainly being on the aesthetic qualities of public space, many of Sitte's observations influenced many subsequent theorists and urban planners.

FURTHER READINGS

Collins, G.R. & Craseman-Collins, C. (2006) *Camillo Sitte: The Birth of Modern City Planning*, Dover Publications: New York.

Sitte, C. & Stewart, C.T. (2013) *The Art of Building Cities: City Building According to Its Artistic Fundamentals,* Reprint of the 1945 edition, Martino Fine Books: Connecticut.

PATRICK GEDDES

Patrick Geddes (1854–1932) was a Scottish biologist and sociologist who is best known for being referred to as the founder of the modern town planning. In 1880, Patrick Geddes was appointed as an academic in botany at Edinburgh University. Over the following

20 years, he undertook many of social experiments designed to improve housing and living conditions in Edinburgh Old Town, which, at the time, needed significant improvements to its squalid housing stock and poor sanitation. He strongly believed that to fully understand and improve the community, he had to live amongst the residents and so he moved his family into the Old Town. Initially, he improved the building in which he lived, but gradually began to encourage his neighbours into communal action (Welter et al. 2000). He also reclaimed disused and derelict spaces to create green spaces and gardens for the local residents. Geddes' aim was to encourage a diverse mixture of people to settle in the area to ensure a vibrant community.

Geddes was a pioneer in terms of understanding that town planning was about much more than the creation of physical places and that its focus should be on people and communities. Indeed, drawing upon his biology background, he argued that the city was not a machine but a complex organism. His ethos was that cities should be inclusive places that are healthy, culturally rich, and environmentally friendly. Geddes also strongly advocated sensitive and financially prudent revitalisation of urban areas through the repair and adaptation of existing buildings and spaces rather than redevelopment. To further enhance the health and welfare of city residents, he also proposed that city's contain interconnecting parks, and streets lined with trees and greenery. Geddes further developed his innovative philosophy of urban planning and articulated this through his books *City Development* (1904) and *Cities in Evolution* (1915).

An innovator in place-based planning and citizen participation, Geddes' impact was international with globally significant contributions to planning theory. He also engaged in landscape and urban development projects around the world (Young & Clavel 2017). Indeed, he was also involved in the planning of Tel Aviv, Bombay, and many other Indian cities, where he focused on the relationship between inhabitants and their surrounding environment. In 1919, Geddes was also commissioned to suggest improvements to the city of Jerusalem, and in 1920, he prepared a report on the re-planning of Colombo, Sri Lanka. Significantly, Geddes' work enabled a deeper understanding of people and their relationships with natural, built, and social environments. Despite much resistance at the time, from engineers focused on redeveloping cities, his ideas and concerns about the environment, and conservation influenced many later theorists and practitioners. Significantly, however, his ideology of community engagement in the revitalisation of urban communities remains pertinent to the challenges facing cities in the twenty-first century.

FURTHER READINGS

Geddes, P. (2019) *Cities in Evolution: An Introduction to the Town Planning Movement and to the Study of Civics*, Forgotten Books: London.

Young, R. & Clavel, P. (2017) Planning living cities: Patrick Geddes' legacy in the new millennium, *Landscape and Urban Planning*, Special Issue, October, Vol. 166. www.sciencedirect.com/journal/landscape-and-urban-planning/vol/166/suppl/C

JANE JACOBS

Despite not having an education or practice experience in any built environment discipline, American journalist and author Jane Jacobs (1916–2006) has had a significant impact upon contemporary urban design concepts. Living and working in Greenwich Village, Manhattan during the 1950s and 1960s, she witnessed how the communities of New York's older city neighbourhoods were being severely and negatively impacted by slum clearance, urban regeneration, and new highway projects under the 1949 Housing Act. In the early 1950s, whilst working for the journal *Architectural Forum*, Jacobs had the opportunity to propound many of her ideas about the injustices and mistakes being made by these urban renewal projects. Subsequently, Jacobs wrote her seminal book *The Death and Life of Great American Cities* (1961) in which she not only attacked the pervading approach to city planning but also articulated her thoughts, and theories for how community-focused cities should be designed and revitalised.

The Death and Life of Great American Cities (1961: 5) commences by stating that: *"This book is an attack on current city planning and rebuilding... on the principles and aims that have shaped modern, orthodox city planning."* The book was written based on her experiences of living amongst the communities being severely impacted by urban renewal projects. Jacobs had a passionate interest in the physical, social, and economic dynamics of the city and this helped her to see the city and its built environment like a naturalist. Indeed, Jacobs (1961) argues that people should rely on *"eyes and instincts"* and to *"seek the truth from the facts"* rather than from maps and statistics. In doing so, she demonstrated how the social interactions of an active pedestrian street life are the essence of urban places. She observed how the traditional neighbourhoods, described as slums by politicians, and planning professionals, were places where communities and small local businesses flourished. Indeed, despite their physical appearance, she saw them as places of vitality, diversity, opportunity, equity, and safety. In contrast, Jacobs identified how the redevelopment projects were typified by over-scaled mono-functional housing blocks with no diversity in terms of tenure, and wide roads with no shops and employment opportunities, no sense of community, and a lack of character.

Jacobs' overriding ambition was to see greater and impactful engagement of local communities in a meaningful participatory process for decision-making for their neighbourhoods. Having seen how the slum clearance programmes were destroying communities, diversity, history, economic, and social opportunities, she became a powerful activist against such developments and became a local hero for her campaigning against these projects. Her opposition to one project in particular, a proposal for a ten-lane highway that would run through the middle of Washington Square Park, drew unprecedented political attention and led to the FBI investigating her. Jacobs' fight against the renewal projects led to New York's chief City Planner, Robert Moses, known as the 'master builder', becoming her infamous adversary (Tyrnauer 2017).

In addition to highlighting the important and inclusive qualities of the existing neighbourhoods, Jacobs also described in detail the townscape qualities that she considered vital to a successful city (Jacobs 1961). Whilst she did not invent all of the concepts, ideas, and theories in *The Death and Life of Great American Cities*, she did announce them to a wider audience through her detailed examination of their importance in the context of real places. Most of these qualities of successful urban environments that Jacobs identified are now commonplace in contemporary urban design guidelines:

- Higher densities of people living in urban areas create a critical mass that can promote vitality, support the local economy, justify public transportation, and provide natural surveillance.
- Streets that are frequent and create small-scale urban blocks. Jacobs noticed how these created permeability and choice in terms of movement around the neighbourhood, increased walkability, and facilitated the interaction of people, identifying that: *"...frequent streets and short blocks are valuable because of the fabric of intricate cross-use that they permit among users of a city neighbourhood."*
- Street corners that act as focal points, landmarks, places of activity, and local meeting places for the local community.
- Streets with active edges and buildings that are oriented towards the street to provide activity and help to ensure 'eyes on the street' and therefore a feeling of safety through strong natural surveillance. Jacobs (1961: 45) stated that: *"...the sidewalk must have users on it fairly continuously, both to add to the number of effective eyes on the street and to induce the people in buildings along the street to watch the sidewalks in sufficient numbers."*
- Public and private spaces should be clearly demarcated and obvious to a city's users in order to avoid confusion.
- Mixed-use buildings and neighbourhoods containing a variety of uses and functions within close proximity to provide choice, diversity, and variety. She identified how the *"intricate minglings of different uses in cities are not a form of chaos. On the contrary, they represent a complex and highly developed form of order."* (Jacobs 1961: 222).
- Varied buildings in terms of age, tenure, and physical condition to create varied residential and business opportunities through varying levels of affordability. She states that: *"The district must mingle buildings that vary in age and condition, including a good proportion of old ones so that they vary in economic yield they must produce. This mingling must be fairly close-grained."* (Jacobs 1961: 187).
- Incremental change and growth in terms of the physical, social, and economical qualities is important in successful and liveable neighbourhoods. This she noted, tended to evolve incrementally and foster a sense of community and feeling of belonging that is difficult to create in large-scale new development projects.

Despite being over 60 years old, *The Death and Life of Great American Cities* remains in publication and has been translated into most major languages and is a core text for urban planning and urban design academic programmes around the World. The significance and impact of the book is emphasised by Peter Laurence in

Becoming Jane Jacobs (Page & Mennel 2011: 15), when he states that: *"Jacobs dealt an epic blow to a multi-billion dollar regime of federal and local policies, agencies and real estate development interests; articulated the bankruptcy of prevailing city planning theories; and wrote one of the most important books on cities and city life."* Indeed, Jacobs' observations led to an about-turn in thinking about cities and planning in the latter half of the twentieth century, and her legacy continues to influence the theory and practice of urban design.

Despite bringing ordinary people, communities, and places into urban development discussions, and highlighting the importance of understanding their concerns and issues, Jacobs has, however, received criticism for her views of the city. Indeed, she has been described as utopian, romantic, and simplistic and whilst her observations and efforts were able to save the townscape and architectural heritage of Manhattan neighbourhoods she failed to save its vitality and the long-term future of the original communities. Indeed, Zukin (2010) discusses how Jacobs helped to create a perfect environment for gentrification to occur and Moskowitz (2016) in describing many of Jacobs' flawed ideas argues that: *"The same neighborhood Jacobs lauded for its diversity in the 1960s and '70s is today a nearly all-white, aesthetically suburban playground for the rich."* Nevertheless, it should be remembered that in the late 1950s and early 1960s, those with choices had abandoned the older city neighbourhoods for the suburbs and there was little sign of them returning. Bratishenko (2016) also suggests that Jacobs's seemingly attractive opinions are not universally applicable once you consider class, ethnicity, race, and other less transparent relationships of power. Jacobs' legacy does however live-on, and shortly before her death in 2005, along with a group of urbanists and activists she established *The Center for the Living City*. Its aim is to build on her ideas and to spread the understanding of contemporary urban life, inspire civic engagement, and to develop creative responses to advance social, economic, and environmental justice (see: www.centerforthelivingcity.org). Jacobs' legacy also lives on through the Jane Jacobs Walk, which is a series of annual free neighbourhood walking and cycling tours, that take place in cities around the world, that aim to connect people with their environment and others who live in their communities.

FURTHER READINGS

Jacobs, J. (1961) *The Death and Life of Great American Cities*, Random House: New York.

Kanigel, R. (2017) *Eyes on the Street: the Life of Jane Jacobs,* Vintage Books: New York.

Laurence, P. (2016) *Becoming Jane Jacobs,* University of Pennsylvania Press: Philadelphia.

Schubert, D. (ed) (2014) *Contemporary Perspectives on Jane Jacobs: Reassessing the Impacts of an Urban Visionary,* Ashgate: Farnham.

Tyrnauer, M. (2017) *Citizen Jane: Battle for the City,* Altimeter Films: New York.

GORDON CULLEN

British architect, urban planner, and author, Gordon Cullen (1914–1994), was influential in establishing the townscape movement, which emerged in response to modernist city planning after the Second World War. Like Jane Jacobs, Cullen was a great observer of cities; however, his focus was on how people experience urban environments and the impact on their senses. In doing so, he examined how settlements had grown organically and the visual richness that resulted from this. This enabled him to develop an innovative approach to urban visual analysis based on human perceptions of time and space alongside the need for visual stimulation. Opposing the impacts of Modernism on cities, Cullen's aim was to highlight the visual qualities and variety of traditional towns and cities. To do this, he focused on the sensual stimulation of experiencing cities that display varied townscape and architectural characteristics. He also emphasised the important qualities of cities that have grown organically and contain buildings with a mix of sizes and age in contrast to monotonous post-WWII city planning (Gosling 1996).

Cullen presented his ideas in *Townscape* (1961) with later editions published as *The Concise Townscape*. In the book, he analyses the buildings and public realms of cities through their physical and visual qualities which he then explains through serial vision illustrations to show how urban spaces change as people progress through them. Serial vision has become an important urban design tool for understanding changing perspectives and sequential views experienced by people as they move through urban spaces. Indeed, Cullen was a pioneer in terms of demonstrating the importance of understanding how people perceive, interpret, and react to urban places. He did this by highlighting three important ways in which people process the visual stimuli of townscapes: (i) visual dimension or how people see the urban environment which he demonstrates through a series of sequential views of existing cities; (ii) the place itself and how people feel and make decisions based on townscape cues such as key views; (iii) the variety of the physical elements, such as scale, mass, style, materials, texture, and colour, and how these help to create identity and stimulate interest (Cullen 1961).

In discussing the importance of placemaking, Cullen (1961) discusses the '*art of relationships*' and the importance of understanding and responding to the emotions and experiences of people in urban environments. He argues that this in turn needs to inform design approaches in order to create visual interest and excitement. Cullen's work presented a new innovative way of seeing and understanding the city, and his ideas remain influential in analysing the character of the built environment from a human perspective. Similarly, his representational techniques such as serial vision continue to be used as both analytical and design tools by urban designers.

FURTHER READINGS

Cullen, G. (1961) *The Concise Townscape*, Architectural Press: Oxford.
Gosling, D. (1996) *Gordon Cullen: Visions of Urban Design*, Academy Editions: London.

KEVIN LYNCH

American urban planner and author Kevin Lynch (1918–1984) who undertook pioneering empirical research on how people perceive and navigate urban environments. The outcomes of his observational and insightful work are presented in two seminal books, *The Image of the City* (1960) and *A Theory of Good City Form* (1984). A central premise of Lynch's research was that people and their perceptions should be a prime consideration when designing urban places. His research therefore focused on how people understand urban environments and one of the innovations of Lynch's research was the use of city residents to record their movements and experiences of their urban environment. The participants were required to produce mental maps of the city to enable him to understand the impact of the city's physical characteristics and visual stimuli. Lynch then analysed these places in relation to their legibility and subsequently invented the term 'imageability'.

In contrast to Jane Jacobs, Lynch studied the city from a more objective rather than personal point of view. In doing so, he articulated the city from two perspectives. The first uses a rational model to define the city as a hierarchy of spaces, whilst the second is based upon how people navigate urban environments in relation to their memory and physical characteristics. In *The Image of the City* (1960: 5), Lynch argues that a city's success as a legible urban form comes from its ability to create a *"vivid and sharp image,"* which in turn becomes part of peoples' collective memory. Further to his research experiments, Lynch (1960) argued that a city's residents process information in consistent and predictable ways by creating mental maps containing five key components: *paths* (the routes through the urban environment); *edges* (boundaries that restrict movement); *districts* (areas with common characteristics); *nodes* (places for orientation such as junctions); and *landmarks* (recognisable objects).

In addition to making a significant contribution to empirical studies and theories on urban form, Lynch's work also advanced the understanding of environmental psychology particularly in relation to the concept of genius loci. Derived from Roman mythology, this refers to the protective spirit of a place and is often now referred to as a sense of place, which was first mentioned in relation to landscape design by Alexander Pope (1688–1744) in the early eighteenth century when writing a series of poems themed around this concept. The value of Lynch's research has drawn some criticism because it focused on spatial function and identity but ignored meanings and experiences. Others have been critical of the potential influence on participants in his research through his instructions and also because peoples' understanding and perception of place will vary considerably based on their age, gender, cultural background, social status, etc. Nevertheless, Lynch's work inspired a considerable number of studies on environmental psychology and behaviour in cities and still impacts upon how designers identify the character of a particular area and design urban environments. Indeed, Lynch highlighted the importance of understanding how people create meaning from places how what is really important is understanding the experience and memories of people living in a particular place.

FURTHER READINGS

Lynch, K. (1960) *The Image of the City*, MIT Press: Cambridge, Mass.
Lynch, K. (1984) *A Theory of Good City Form*, MIT Press: Cambridge, Mass.
Salerno, R. (2014) Rethinking Kevin Lynch's Lesson in Mapping Today's City, pp. 25–31, In: Contin, A., Paolini, P. & Salerno, R. (eds.) *Innovative Technologies in Urban Mapping*, Built Space and Mental Space, Springer: New York.

WILLIAM H. WHYTE

As a sociologist and urbanist, William H. Whyte (1917–1999) was a pioneer of research into human behaviour in the public realm. Working for the New York City Planning Commission in the late 1960s, Whyte identified the need to understand how people were actually using the city's public spaces. This led to *The Street Life Project* in which he studied pedestrian behaviour in the New York's streets, parks, and other public spaces. Whyte presented some of the results of the project in his book *The Social Life of Small Urban Spaces* (1980), and a film with the same title. His research focused on understanding who was using the city's public spaces and how they were using these places based on covert observational studies. In addition to direct observation, Whyte used time-lapse photography and video recordings to provide evidence-based evaluations of the use of public space. This enabled him to challenge existing theories whilst also informing planning policy and public realm design.

Through his research and publications, Whyte emphasised how the positive impact that social life in the public realm can contribute towards people's quality of life in towns and cities. Indeed, Whyte (1988: 109) eloquently articulated that: *"It's hard to design a space that will not attract people. What is remarkable is how often this has been accomplished."* He also emphasised the need for urban spaces that enabled civic engagement and facilitated the engagement of communities in a wide range of activities. To enable the design and creation of such spaces, Whyte firmly believed in the need for a bottom-up approach to design that was informed by an in-depth understanding of how people use and how they would like to use public space. His empirical data as a result of detailed observations and talking to the users of public spaces enabled Whyte to significantly advanced knowledge and understanding of who, how, and why people use public places within our cities. In turn, Whyte's work continues to play a critical role in informing contemporary urban design practice. Significantly, Whyte was the mentor for the US-based Project for Public Spaces because of his seminal work in the study of human behaviour in the public realm. Indeed, Project for Public Spaces founder and president Fred Kent worked as one of Whyte's research assistants he based the organisation largely on Whyte's methods and findings (PPS 2024).

FURTHER READINGS

LaFarge, A. (ed.) (2000) *The Essential William H. Whyte*, Fordham University Press: New York.

Project for Public Spaces (2024) *William H. Whyte: Placemaking Heroes.* www.pps.org/article/wwhyte

Rein, R.K. (2022) *American Urbanist: How William H. Whyte's Unconventional Wisdom Reshaped Public Life*, Island Press: Washington DC.

Whyte, W.H. (1988) *City: Rediscovering the Center*, University of Pennsylvania Press: Philadelphia.

Whyte, W.H. (2021) *The Social Life of Small Urban Spaces*, 8th edition, Project for Public Spaces: New York.

CHRISTIAN NORBERG-SCHULZ

Norwegian architect and theorist, Christian Norberg-Schulz (1926–2000), expanded on the concept of genius loci in his highly influential works, such as *Intentions in Architecture* (1963), *Existence, Space and Architecture* (1971), and *Genius Loci: Towards a Phenomenology of Architecture* (1979). Expanding upon Lynch's focus on how physical elements enable people to understand urban areas, Norberg-Schulz related understanding of the meaning to places such as how a space can have a character and what that character means for people and *"how people subjectively and intersubjectively holistically experience places with all their senses"* (Portugali 2024). In doing so, he focused on the phenomenology and psychology of place and in order to articulate the significance of meaning and experience when inhabiting places in the real world. Norberg-Schulz argued that places are connected to history and community and that physical spaces possess socio-cultural significance and an identity of their own. In his work, he focuses upon the relationship between architecture, place, and identity in order to emphasise that the spirit of the place, or genius loci, is central to understanding the built environment.

In *Intentions in Architecture*, Norberg-Schulz examined theories around the organisation of built form and space in order to highlight the importance of visual perception. He also stressed the importance of place to peoples' existence and that as such, place is not an abstract or quantifiable but a qualitative phenomenon. This enabled him to develop a method of phenomenological analysing cities that he later articulated in *Genius Loci: Towards a Phenomenology of Architecture*. In this, he argued that space in an urban context is intangible from a psychological perspective since it is derived from a users' sensory experience. Norberg-Schulz identified that a sense of place can evoke feelings of comfort, satisfaction, and security for a city's inhabitants and that this can positively impact upon their quality of life. He further argued that objective and scientific approaches couldn't provide an understanding of how people meaningfully experience the environments they engage with. Indeed, he stated that spaces where life occurs are places, and places are spaces with a distinct character or genius loci. Norberg-Schulz's theories that places embodied unique characteristics and meanings that could not be understood purely from analytical or scientific analysis had considerable influence upon later theorists and practitioners.

FURTHER READINGS

Norberg-Schulz, C. (1979) *Genius Loci: Towards a Phenomenology of Architecture*, Rizzoli: New York.

Norberg-Schulz, C. (1988) *Architecture, Meaning and Place: Selected Essays* Rizzoli: New York.

CHRISTOPHER ALEXANDER

Christopher Alexander (1936–2022) was a Viennese-born architect, theorist, and professor at the University of California, Berkeley. Alexander was a fierce anti-modernist who argued that traditional buildings and settlements were more aesthetically pleasing than modern ones and that the most beautiful and successful places were not designed by architects but made by the people. Alexander's ideas focused on the ways in which cities were organised and designed to ensure life could flourish, arguing that answers for contemporary design could be found in successful places from the past. Indeed, he argued that successful urban areas were living frameworks for human beings, particular those that had developed organically into places that positively support people's lives.

In his essays entitled *A City is Not a Tree* (1965), Alexander argued against the simplistic structures of planned cities and raised profound questions about how cities should be designed. He identified how traditional cities had a characteristic overlapping web–network structure, which emerged as a spontaneous outcome to serve a city's inhabitants and that the outcomes of this process could often be seemingly mundane. Indeed, he gave examples of simple urban elements such as a sidewalk, a crosswalk light, and a newspaper rack, suggesting that these form part of a complex interactive network. In contrast, he considered that town planners had been rationally organising cities into neat, tree-like geometric forms, with profoundly negative consequences for the life of the city. Alexander considered that the separation of different functions that was dominating urban planning at the time was an unhealthy way of building cities. Innovative at the time, and drawing upon his observations of historic cities, he argued that for a city to be receptive to life, social interaction, and human prosperity, it must contain a mix of functions and be allowed to change organically. Significantly, he also understood that whilst the relationship between pedestrians and cars was challenging, that their simple separation is not a sustainable solution for making cities liveable.

As a result, he suggested a pattern for analysing and improving the interactions between vehicles and people whereby their interactions can be safe, mutually beneficial, and vibrant.

Alexander subsequently published a number of important books, including *A Pattern Language: Towns, Buildings, Construction* (1977), The Timeless Way of Building (1979), A New Theory of Urban Design (1987), The Nature of Order (2002), and The Battle for the Life and Beauty of the Earth (2013), that further expanded his theoretical stance. In *A Pattern Language*, one of Alexander's most eminent publications, he describes 253 'patterns' or qualities that could be used to design a home, a neighbourhood, or a city. Having observed the alienating impact of poor architecture and city design, Alexander set out to establish a guide through the language of patterns that would enable the design and making of almost any kind of building, or any part of the built environment. Drawing upon his observations of medieval cities and how individuals were able to change and adapt their environment to meet their needs, he developed the patterns as tools to *"allow anyone, and any group of people, to create beautiful, functional, and meaningful places."* One common criticism of Alexander's language of patterns is that it was too restrictive; however, a strength of the patterns was that they were intended to be revised by its users and be inherently flexible. Significantly, Alexander's work has influenced a large and diverse group of respected architects, urban planners, researchers, and theorists, including the New Urbanism movement (Mehaffy 2008).

FURTHER READINGS

Alexander C. (1965) *A City is Not a Tree, Architectural Forum*, 122(1): 58–62 (Part I) April, and 122(2): 58–62 (Part II) May.

Alexander, C., Ishikawa, S., Silverstein, M., Jacobson, M., Fiksdahl-King, I. and Angel, S. (1977) *A Pattern Language: Towns, Buildings, Construction*, Oxford University Press: New York.

Alexander, C., Neis, H., Anninou, A. & King, I. (1987) *A New Theory of Urban Design*, Oxford University Press: New York.

Mehaffy, M.W. (2008) Generative Methods in Urban Design: A Progress Assessment, *Journal of Urbanism*, 1(1): 57–75, https://doi.org/10.1080/17549170801903678

JAN GEHL

Danish urban designer and author Jan Gehl (1936–) is arguably the most eminent urban design practitioner of the late twentieth and early twenty-first century. Not only has Gehl positively influenced the transformation of cities around the World, but his influential research has also significantly enriched knowledge into how inhabitants actually use the urban environment. Upon graduating in architecture from the Royal Danish Academy of Fine Arts in 1960, Gehl worked as an architect; however, he also credits his wife, a psychologist, for helping him to understand the human side of architecture through their discussions on the relationships between psychology and sociology and the way that people use and experience cities. Significantly, Gehl was able to test his ideas on urban life further when in 1966 he was awarded a research grant to study the form and use of public spaces. The project enabled him to develop his theories on how architecture and city planning influences public life, with findings published in his influential book *Life Between Buildings* (1971).

In *Life Between Buildings*, Gehl emphasised the critical importance of understanding the human dimension of urban spaces including how and why people use public space by studying the human body and senses. He identified these as the location of social interaction and sensory experience and in doing so argued that people's outdoor activities can be classified into three categories:

- *Necessary activities* are the everyday tasks that have to be undertaken and include those such as walking to work, going shopping, taking children to school, and walking the dog. The quality of the environment has little influence over these activities as they must happen regardless of the weather, and the conditions of the urban environment.
- *Optional activities* are likely to be more dependent on the quality of the environment and only tend to occur when there is a desire to do them, the weather is reasonable, and there is the time, and place for them to occur. Unlike necessary activities, optional activities are more likely to occur, as people have a choice, in places with favourable physical characteristics. Optional activities include things like walking for leisure, sitting outside a café, playing sports or games, people watching, etc.

- *Social activities* are an important dimension of the animation of public spaces, and these are prevalent when people congregate and socialise. Social activities often tend to be spontaneous and result from people being together and in close proximity. These can include things like passers-by acknowledging each other, children meeting and playing, friends chatting, overhearing conversations, sharing a newspaper, smiling at a stranger, etc.

Gehl identifies that the better the quality of the place, the more optional activities will occur and any necessary activities will last longer. He also suggests that social activities tend to result from people taking part in necessary or optional activities, but like optional activities, they are more likely to occur in better-quality urban environments. As a result, more successful public places are likely to facilitate more optional and social activities, which in turn will make them even more attractive to users. In contrast, poorer quality urban environments are unlikely to witness anything other than necessary activities, and even then, only when people have no convenient alternative choices.

Gehl's approach to improving or designing public spaces led to a number of other significant publications, such as *Public Spaces – Public Life* (1996), *New City Spaces* (2008), *Cities for People* (2010), and *How to Study Public Life: Methods in Urban Design* (2013) further develop his methods. In *Cities for People*, Gehl built upon his earlier work to emphasise the importance of designing to a human scale and the essential nature of walkability in creating successful urban spaces.

Significantly, his work focuses on how to 'see' and 'read' urban spaces, and through these publications, he explains: (i) how to undertake a detailed evaluation the quality of public places by recording public life; (ii) how to facilitate the active use of urban spaces; and (iii) how people's experience and perception affects their use of the public realm. He then demonstrates how to use this research to inform design solutions and recommendations for the creation of useable places. Indeed, collecting and analysing empirical data is critical in his approach in terms of ensuring contextually appropriate solutions, but also for educating and convincing the key decision-makers such as planners and politicians.

Much of Gehl's early research focused upon his home city of Copenhagen and his ideas and theories have since been instrumental in helping the city to shape its identity as one of the World's most walkable cities. Such was the success of his initial work that this quickly expanded to have global influence and his practice Gehl Architects have applied their urban design principles to cities throughout Europe, Asia, Australia, and North America. The practice's work is informed by a deep understanding of people and places resulting from years of research into cities and public life. Indeed, Gehl has been highly influential as a practitioner, researcher, educator, and author and has had a direct impact on the design and enhancement of cities whilst furthering the understanding of what makes a people-friendly city.

FURTHER READINGS

Dalsgaard, A. (2012) *The Human Scale: Bringing Cities to Life,* Final Cut for Real: Copenhagen.

Gehl, J. (1996) *Public Spaces – Public Life*, Island Press: Washington DC.

Gehl, J. (2010) *Cities for People*, Island Press: Washington DC.

Gehl, J. (2011) *Life Between Buildings: Using Public Space*, 6th edition, Island Press: Washington DC.

Gehl, J. & Gemzøe, L. (2008) *New City Spaces*, Danish Architectural Press: Copenhagen.

Gehl, J. & Svarre, B. (2013) *How to Study Public Life: Methods in Urban Design*, Island Press: Washington DC.

Gehl People (2024) www.gehlpeople.com

Matan, A. & Newman, P. (2017) *People Cities: The Life and Legacy of Jan Gehl*, Island Press: Washington DC.

PROJECT FOR PUBLIC SPACES

The Project for Public Spaces is a non-profit organisation that focuses on enabling people to create and sustain public places that help to build resilient communities. Project for Public Spaces was founded in 1975 by Fred Kent who was one of William H. Whyte's research assistants. Building upon Whyte's research, the Project for Public Spaces aims to transform and invigorate public spaces by designing them in collaboration with the people and communities who will use them. Project for Public Spaces (2023a) has engaged in this approach to the redesign and reinvigoration of public spaces through community-based placemaking to *"inspire people to collectively reimagine and reinvent public spaces at the heart of every community."* Indeed, in all of its projects, it stresses that a collaborative approach is imperative and that there should be a focus on all of the identities of a place if its on-going evolution is to be ensured.

In order to identify what makes a successful public space, Project for Public Spaces (2023b) developed *The Place Diagram* as a tool to enable the qualitative and quantitative evaluation of public places. Based on their experience of evaluating public spaces, the diagram identifies four key qualities that successful places generally display: they are *accessible*; people engage in *activities*; they are *comfortable* and have a positive image; and they are *sociable* places. The Place Diagram uses these four fundamental qualities – uses and activities, comfort and image, access and linkages, and sociability – with sub-categories to set a series of questions that enables the relative successes or failures of a space to be evaluated. Building upon its implementation of placemaking projects, Project for Public Spaces developed guidelines for public space revitalisation based on *11 Principles of Placemaking* (Madden 2021). Published in *How to Turn a Place Around* (1999), Project for Public Spaces (2023c) identified eleven key principles that can transform public spaces into vibrant and animated places. These are intended to support communities to: (i) integrate diverse opinions into a cohesive vision; (ii) translate the vision into a plan and programme of uses; and (iii) ensure the plan's sustainable implementation.

Emphasising the paramount importance of placemaking, Project for Public Spaces (2024) define it as: *"a multifaceted approach to the planning, design, and management of public spaces."* They further elaborate by defining the key characteristics and considerations of placemaking: community-driven; visionary; function before form; adaptable; inclusive; focused on creating destinations; context-specific; dynamic; transdisciplinary; transformative; flexible; collaborative; and sociable. To further evaluate and facilitate placemaking, Project for Public Spaces (2022) developed the *Power of 10+* concept. This focuses on the importance of considering human scale in projects and the significant impacts that this can have on the success of a city's public places. The idea behind the concept is that places will flourish when users have a whole range of reasons and motivations (more than ten) to be there and to use the public realm. This can include a whole range of activities, such as places where there are opportunities to sit, relax, eat, meet others, people watch, exercise, allow children to play, play games, sunbathe, read, walk dogs, listen to music, watch performances, admire artwork, buy things, and on and on (PPS 2023d). It is acknowledged that not all places will facilitate all of these activities; however, the more opportunities there are the better. Also, ideally some of the activities will be unique to the place and be inspired by its particular and unique contexts whether they be cultural, social, historical, physical, etc. Importantly, Project for Public Spaces stresses the critical importance of engaging local communities because they will know which specific activities are most appropriate and therefore likely to be successful. PPS goes further by suggesting that all cities should contain at least 10 destinations containing at least 10 such places, with each in turn have over 10 things for people to participate in.

Significantly, although US-based, Project for Public Spaces, through its research, publications, education programmes, partnerships, and technical assistance, has had a considerable impact upon the realisation of successful public places in partnership with local communities, not just in the United States but also in over fifty other countries (PPS 2024) together with influencing practitioners around the world.

FURTHER READINGS

Madden, K. & Project for Public Places (2021) *How to Turn a Place Around: A Placemaking Handbook,* PPS: New York.

Project for Public Spaces (2022) *Placemaking: What If We Built Our Cities Around Places?,* PPS: New York.

Project for Public Spaces (2024) www.pps.org

SUMMARY

This chapter has focused on the key figures and organisations behind the evolution of the theories that underpin urban design as a discipline. In doing so, the changing approaches and attitudes to place and placemaking have been examined through the ideas, philosophies, and works of renowned theorists, academics, organisations, and practitioners. The chapter has articulated the increasing concern with people as part of human-centric urban design since the late nineteenth century has changed the focus of how we see and understand the public realm. Indeed, the intrinsic relationships between place and people alongside the importance of contextual responses and sustainability are now at the forefront of urban design thinking. This knowledge is important for urban designers in terms of furthering their understanding of the public realm from a theoretical through to a practice-based standpoint. In turn, this enables the urban design profession to position its own approaches to practice in relation to critical thinking about approaches to placemaking.

4

COMPONENTS AND PRINCIPLES OF URBAN DESIGN

Although the urban environments of most towns and cities have evolved over centuries, a successful public realm is increasingly seen as a fundamental component of a vibrant and sustainable settlement. Urban design has, therefore, a primary role to play in ensuring that the public spaces in urban areas are designed and managed in a contextually sensitive way. If done successfully, then towns and cities around the world can fulfil the needs of the people and communities that inhabit and visit them. Indeed, Raynsford and Lipton (DETR 2000: 5) suggest that: *"Good urban design is essential if we are to produce attractive, high-quality, sustainable places in which people will want to live, work and relax."* To achieve this, it is imperative that urban designers understand what common qualities help to make a successful public space and how these can be achieved in different urban contexts. This involves not just an in-depth knowledge of the main physical components of urban design – the street, the public square, and the urban block – but also the basic principles that apply to the design and success of these elements. Indeed, well-designed and well-maintained public spaces are essential to the success of all urban environments. This chapter will therefore focus on defining the qualities that constitute a successful public space, the fundamental components of urban design, and the basic principles that help to ensure high-quality urban design outcomes.

DOI: 10.4324/9781003251200-4

SUCCESSFUL PUBLIC SPACES

Public spaces have existed and fulfilled vital roles in settlements around the world for thousands of years. The streets, squares, and public parks that make up the public realm are not a recent invention with their being endless examples of beautiful and functional public spaces that successfully meet the needs of people using towns and cities. Indeed, urban environments contain a plethora of public places and everyday spaces such as streets, squares, parks, waterfronts, markets, arcades, bridges, and public buildings that are freely accessible to the public. It is important to note, however, that not only will urban areas and the public spaces within them evolve over time, so will the way in the which the public uses them because of changing demands and expectations. As such, whilst there are countless examples of successful streets and squares throughout history, it is imperative that urban designers can recognise the value of the successful historic qualities of public places but be able to reconcile these with contemporary requirements. Indeed, the traditional roles of streets and squares as places for circulation, trading, and festivals have been surpassed by an infinite number of functions and activities in contemporary society (see also the section on Mixed-Use and Multifunctional).

What makes a successful public place is a continually evolving challenge; however, there are many accepted qualities that should be considered during all stages of the urban design process and the on-going management of the spaces. In simple terms, successful public spaces are those places that people make a positive decision to use. In practice, these are likely to be places where social and economic exchanges occur, friends meet, cultures mix, and places where celebrations and events occur. In contemporary society, public places should contain as many reasons as possible for people to want to be there. Obviously, cultural expectations will vary around the world; however, there should be many things for people to do within public spaces including sometimes unexpected opportunities that result in people staying for longer in the space. To facilitate this, urban design and placemaking can and should play a fundamental role in the design, refurbishment, use, and management of public spaces. Indeed, despite their different contexts and purposes, there are many common qualities that all these

Figure 4.1 Successful public streets create pedestrian-friendly environments that are inclusive and provide many different reasons for people to use them (Qianmen Street, Beijing, China).

Source: Photograph taken by the authors.

spaces should attempt to embody if they are to be successful. It is important to note, however, that whilst all these qualities play an important role in the achievement of successful public spaces, it is not always important that specific places display all these qualities and that many successful public places will only display some of these (Figures 4.1 and 4.2).

Many practitioners, organisations, and researchers have attempted to define what makes a successful public space and how to undertake effective placemaking (Gehl 2010; Madden & Project for Public Spaces 2021; Project for Public Spaces 2022; Carmona 2021; Chicago MPC 2008; Essex County Council 2021). Whilst it is commonly acknowledged that high-quality public spaces can come in all shapes and sizes, have many different purposes and functions, contain an infinite range of activities, exist in varied cultural and climatic contexts, etc., there are many key qualities of good public spaces that help to ensure their success. Project for Public Spaces (2008), in discussing the qualities of great streets, identifies

Figure 4.2 Successful public squares attract a wide range of users for a variety of different activities (Kultorvet Square, Copenhagen, Denmark).

Source: Photograph taken by the authors.

ten key considerations: attractions and destinations; identity and image; active edge uses; amenities; management; seasonal strategies; diverse user groups; traffic, transit, and the pedestrian, blending of uses and modes; protects neighbourhoods. More specifically in relation to all public places, Project for Public Spaces (2005), through their research, also developed *The Place Diagram tool*, which identifies that most great public spaces share four fundamental qualities and poses a series of questions that can be used to assess existing and proposed places:

1. Access & Linkages: convenient, accessible, and well connected, both visually and physically, to other important places in the area;
2. Comfort & Image: comfortable with appropriate facilities whilst presenting a good image and positive perceptions;
3. Uses & Activities: varied attractions that will bring people to the space and encourage them to return;
4. Sociability: sociable place where people interact and having a sense of attachment.

Additionally, many local government authorities have adapted the numerous criteria that define what makes a successful public space in their own design guidelines. *The Essex Design Guide* (Essex County Council 2021), for example, recommend a series of criteria to consider when designing public open space. In considering what can make public places better for people, Jan Gehl (2010), in *Cities for People*, suggests that there are twelve fundamental criteria organised into three categories that should be considered during the design process: (1) *Protection* against: (a) traffic and accidents (feeling safe), (b) crime and violence (feeling secure), and (c) unpleasant sensory experiences; (2) *Comfort* through opportunities to: (a) walk, (b) stand/stay, (c) sit, (d) see, (e) talk and listen, and (f) play and exercise; and (3) *Environmental* considerations: (a) human scale, (b) enjoy positive aspects of climate, and (c) provide positive sensory experiences. Whilst the various criteria and qualities vary slightly, the clear intention of all guidelines is to facilitate a better quality public realm for all of its users.

FURTHER READINGS

Essex County Council (2021) *The Essex Design Guide: Successful Criteria for Public Open Spaces*, Essex County Council: Chelmsford, www.essexdesignguide.co.uk/design-details/landscape-and-greenspaces/successful-criteria-for-public-open-spaces

Gehl, J. (2010) *Cities for People*, Island Press: Washington DC.

Madden, K. & Project for Public Places (2021) *How to Turn a Place Around: A Placemaking Handbook*, PPS: New York.

Ministry of Housing, Communities & Local Government (2021) *National Design Guide: Planning Practice for Beautiful, Enduring and Successful Places*, MHCLG: London, https://assets.publishing.service.gov.uk/government/uploads/system/uploads/attachment_data/file/962113/National_design_guide.pdf (accessed 21 August 2023).

Project for Public Spaces (2005) *What Makes a Successful Place?: The Place Diagram*, www.pps.org/article/grplacefeat

Project for Public Spaces (2022) *Placemaking: What if We Built Our Cities Around Places?: A Placemaking Primer*, PPS: New York.

Sim, D. (2019) *Soft City: Building Density for Everyday Life*, Island Press: Washington.

COMPONENTS OF URBAN DESIGN: STREETS, SQUARES, AND URBAN BLOCKS

From an urban design perspective, there are three fundamental components that constitute the urban form of our towns and cities: the street; the public square; and the urban block. Whilst buildings, public parks and other green spaces, rivers, and other water bodies are also important to the overall appearance and experience of urban environments, it is these three components over which urban designers have the greatest influence. The streets and squares of a town or city are multidimensional and dynamic spaces that people engage with when using the urban environment and in turn experience with all their senses. The urban block is the architectural form that provides not only the three-dimensional enclosure to urban space but also the connection between external space and internal uses and activities. This section will therefore focus on these three principle components of urban design and on the importance of considering regional and global contexts.

THE URBAN STREET

Although they take on many forms, scales, appearances, and purposes around the world, the predominant public realm in all cities is the street. Indeed, streets are the basic organisational component of cities, with Jane Jacobs (1961) describing them as vital organs which play a vital role in city life. As a conduit for movement, interaction, and transaction, streets are the components of cities that make them function whether this be for mobility or for stationary activities, for leisure or for work, out of necessity or by choice. As public spaces, streets therefore help to activate cities both socially and economically. As such, the use of urban streets is incredibly diverse and includes all types of movement and transportation, places for providing services, businesses, and trading, together with recreational activities. The various activities that streets accommodate and facilitate play a central role in shaping the accessibility, character, and liveability of cities. As such, because of the wide range of users, activities, and requirements placed upon streets, it is imperative that they are well designed and maintained to ensure that they

Figure 4.3 Streets can serve many different functions that contribute to their character and qualities (Jalan Hang Kasturi in Kuala Lumpur, Malaysia).
Source: Photograph taken by the authors.

are safe, accessible, and attractive. As such, they are a fundamental and necessary investment for cities, with successful streets being able to significantly enhance the local economy, health of citizens, and maximise social capital with urban areas (Figures 4.3 and 4.4).

From an environmental and inclusivity perspective, it is important that streets are designed or reconfigured to favour more sustainable modes of travel to encourage walking, cycling, and taking public transportation, whilst also giving due consideration to the movement of goods and services through cities (see also the sections on Walkability and Multi-Model Transit). It is also vital to consider the safety and accessibility of streets for all of its diverse users with particular attention being paid to the young, the elderly, and those with disabilities and other needs to ensure that streets are not only safe, comfortable, and welcoming, but also inclusive and equitable environments for everyone. Indeed, different people will use and experience streets in different ways depending on their senses, including vision, hearing, and mobility. It is therefore essential that designers consider how materials, textures, sounds, and visual information can assist in creating a safe and enjoyable street environment for people of all abilities. It is also important,

COMPONENTS AND PRINCIPLES OF URBAN DESIGN 71

Figure 4.4 The activities on a street together with the enclosure created by the building facades help to create its character (Rue Nicolas Flamel, Paris).

Source: Photograph taken by the authors.

therefore, to design from a human perspective with an understanding of how people walking and cycling experience and perceive streets in terms of human eye level and speed of travel.

In addition to the multitude of ways in which people use and experience streets, there are also endless physical permutations in terms of their design, layout, and formality. Indeed, the contrasts not just between towns and cities but within different neighbourhoods and quarters of settlements can be extreme and add unique qualities to the experience of visiting these places. From the narrow and meandering informal streets and lanes of European medieval towns, Middle Eastern neighbourhoods, and South-East Asian cities, to the formal grand formal boulevards and avenues such as those in Paris, Vienna, and Washington D.C., the design possibilities are extensive. In terms of the design qualities of urban streets, there are various characteristics that differentiate different street typologies, including the street section (or width-to-height ratio), the formality or informality of both its layout and adjacent architecture, the length of the street, views along or out of the street, trees and landscaping, uses in the buildings and activities within the street itself, facilities and infrastructure, transportation, etc. It is all of these qualities and characteristics that help to make streets unique and in turn add to their visual appeal and interest (Figures 4.5 and 4.6).

Figure 4.5 Streets can be formal and grand in their appearance resulting from the layout and surrounding buildings (Rua Augusta, Lisbon, Portugal).

Source: Photograph taken by the authors.

Figure 4.6 Streets can be more informal and organic in their layout and character (Steep Hill, Lincoln, UK).

Source: Photograph taken by the authors.

As both pathways for movement, and destinations in their own right, the design of successful streets is fundamental to the role that urban design can play in ensuring a high-quality public realm in urban areas. Well-designed streets can then be places for social and economic interaction, cultural expression, social engagement, play, and celebration. When well designed, streets can therefore play pivotal roles as catalysts for urban transformation and the creation of people-friendly and sustainable cities.

FURTHER READINGS

Global Designing Cities Initiative (2016) *Global Street Design Guide*, Island Press: Washington. https://globaldesigningcities.org/publication/global-street-design-guide

Jacobs, A. (1993) *Great Streets*, Cambridge, Mass: MIT Press.

Kiang, H.C., Liang, L.B. & Limin, H. (eds.) (2010) *On Asian Streets and Public Space,* NUS Press: Singapore.

Kostoff, S. (1992) *The City Assembled: The Elements of Urban Form Through History*, Thames & Hudson: London.

Moughtin, C. (2003) *Urban Design: Street and Square*, Third Edition, Architectural Press: Oxford.

National Association of City Transport Officials (2013) *Urban Design Street Guide,* Island Press: New York.

THE PUBLIC SQUARE

In most parts of the world, public squares have, for hundreds or even thousands of years, been at the heart of towns and cities, serving as focal points for civic activities and celebrations, economic trade, cultural exchange, and social interaction. Over time, the roles of these public spaces have witnessed significant transformations, reflecting the social, economic, political, and lifestyle changes and trends. Historically, they predominantly served as places for religious, ceremonial, or market-focused activities; however, in contemporary society, they tend to host a much broader and changeable range of events. Indeed, the public square has generally become a flexible and multi-functional place that caters for a wide range of social, cultural, and economic activities.

In addition to serving as a central node for a plethora of uses and activities and acting as a stage for formal and informal occasions, the public square also plays a key role in enhancing the visual attractiveness of urban environments. Indeed, historically many researchers have focused on the physical and aesthetic dimensions of public squares and plazas (Zucker 1959; Haneman 1984; Kostof 1992; Moughtin 2003; Wachten & Neubauer 2010; Sitte 2013). Clearly, public squares, like streets, are important elements of city design that come in an endless array of different shapes and sizes that reflect their respective design ideology, era of construction, intended function, prevailing climate, topography, etc. (Moughtin

COMPONENTS AND PRINCIPLES OF URBAN DESIGN 75

Figure 4.7 Public squares can be intimate and informal in shape, scale, and appearance (Neal's Yard, London, UK).

Source: Photograph taken by the authors.

2003). Also, as with streets, public squares can be found in an endless array of shapes and sizes together with degrees of formality and informality (Figures 4.7–4.9).

In trying to simplify the analysis of urban squares, Zucker (1959) described three typologies: formal spaces reinforced by formal buildings; formal spaces contrasted with informal buildings; and informal spaces and buildings. Zucker (1959) further described four different forms of public squares: the closed square where the urban space is self-contained (such as the Place des Vosges in Paris or Plaza Nueva in Bilbao); the nuclear square where the space is formed around a central focal point (such as Cavendish

Figure 4.8 Public squares can be formal and grand in appearance whilst also offering a strong sense of enclosure (Place des Vosges, Paris, France).

Source: Photograph taken by the authors.

Square in London or Piazza del Popolo in Rome); the dominated square where the space is directed towards a dominant building (such as Piazza del Campo in Sienna or Markt Square in Delft); and grouped squares where multiple urban spaces are linked as part of one greater composition (such as Place de Thessalie in Montpellier). In terms of the design characteristics of public squares, there are also, therefore, many characteristics that imbue different physical and experiential qualities. These include its size and scale, the degree of enclosure, the formality or informality of both its layout and adjacent architecture, uses in the buildings and activities in the space, the connectivity of the space to neighbouring streets and other public spaces, the facilities and infrastructure, transportation within the space, topography, the views within, into, and out of the space, trees and landscaping, water features, etc. It is all these qualities and characteristics that help to make public squares important urban components of urban environments (Figures 4.10–4.12).

Public squares also provide a crucial sense of belonging and identity for a city's residents and help to bring communities together

COMPONENTS AND PRINCIPLES OF URBAN DESIGN 77

Figure 4.9 Public squares can come in many different shapes and sizes (Plaza Redonda, Valencia, Spain).

Source: Photograph taken by the authors.

Figure 4.10 Squares that are uncomfortable in scale and have very few reasons for people to use them tend to be less successful (City Hall Plaza, Boston, USA).

Source: Photograph taken by the authors.

Figure 4.11 Successful public squares often have the flexibility to accommodate different activities and events at different times (Old Market Square, Nottingham, UK).

Source: Photograph taken by the authors.

to foster a sense of collective ownership. Indeed, as physical landmarks integral to the urban form of a city, they provide a forum for community interaction, and cultural and artistic expression. As places for markets, festivals, public performances, events, or demonstrations, public squares facilitate the connections, relationships, and exchanges that are central to the social fabric of urban life (Moughtin 2003). The design of public squares, therefore, needs to reflect the anticipated users and occupants; however, it can be more important to create a space that has limitless possibilities in terms of the way in which it might be used. Indeed, designing

Figure 4.12 Public squares should provide reasons for people to use them at different times of the day and in different seasons of the year (Plaça de la Independència, Girona, Spain).

Source: Photograph taken by the authors.

squares as places that have the flexibility to accommodate unanticipated changes in terms of taste, fashion, technology, etc., can help to ensure the sustainability and longevity of the space. As such, whilst a square needs to look and feel complete, it is the use of the space essential that its design supports and fulfils people's current and future needs and requirements whether they be individuals, groups, communities, or tourists and visitors. It is these users, as actors within the space, that will determine its success and imbue the square with a sense of place and symbolic meaning. Whilst most public squares, therefore, have defined physical forms including their shape, enclosure, materials, orientation, boundaries, and so on, to be successful their design needs to consider a host of factors that extend beyond its physical dimensions. Consequently, the design and refurbishment of public squares needs to be human-centred, with the requirements their potential users being the primary consideration in both the design process and their ongoing management.

FURTHER READINGS

Burns, M. (2020) *New Life in Public Squares,* RIBA Publishing: London.
Childs, M.C. (2006) *Squares: A Public Place Design Guide for Urbanists* University of New Mexico Press: Albuquerque.
Corbett, N. (2004) *Transforming Cities: Revival in the Square,* RIBA Enterprises: London.
Kostoff, S. (1992) *The City Assembled: the Elements of Urban Form Through History,* Thames & Hudson: London.
Krier, R. (2006) *Town Spaces: Contemporary Interpretations in Traditional Urbanism,* Birkhäuser: Basel.
Lang, J. & Marshall, N. (2016) *Urban Squares as Places, Links and Displays: Successes and Failures,* Routledge: New York.
Marron, C. (ed.) (2016) *City Squares: Eighteen Writers on the Spirit and Significance of Squares Around the World,* Harper Collins: New York.
Moughtin, C. (2003) *Urban Design: Street and Square,* Third Edition, Architectural Press: Oxford.
Wachten, K. & Neubauer, H. (2010) *Urban Design and Architecture: The 20th Century,* H.F. Ullmann Publishing: Cologne.

THE URBAN BLOCK

The urban block is the interaction of urban design, urban planning, and architecture. Like streets and squares, the urban block is a fundamental physical component of urban forms across the world. The arrangement of buildings with their varied uses and functions into urban blocks has been fundamental to the physical organisation for thousands of years with the block being the main ordering element of Roman towns. It is not, however, part of a city's public realm, yet it is a fundamental consideration for urban design as it plays a critical role in how cities are planned and organised whilst also providing the important edges or facades to the public realm itself. From an urban design perspective, the urban block helps to create an ordered backdrop to the public realm by providing enclosure, strong definition of space, and natural surveillance within towns and cities.

The urban blocks critical importance to urban areas is identified by Tarbatt and Street-Tarbatt (2020: 3) who argue that: *"Without the block, there would be no streets, just roads. Without streets there would be no street life, just traffic. Without street life, there would be no city, just buildings ...it is the nature of the interface between the two that has a critical impact on the quality of the spaces between those buildings."* The urban block, therefore, whilst being an architectural element in the townscape has a synergistic relationship with a city's streets and squares both physically and functionally (see also the section on Active Edges).

Despite its long history, the urban block went out of fashion in the middle of the twentieth century, as modernist planning ideals pursued new radical urban environments (see Chapter 2). From the mid-1960s, however, architects such as Aldo Rossi and the Italian Rationalist School, followed by others including Leon Krier and Rob Krier, began to re-examine historical urban and architectural typologies as precedents to inform contemporary design. Indeed, *The Architecture of the City* (1966) by Rossi was instrumental in resurrecting urban typologies and *Town Spaces: Contemporary Interpretations of Traditional Urbanism* (2006) by Rob Krier further advocated the block as the fundamental component of the urban fabric. Subsequently, the urban block, with its inherent qualities, once again became the inspiration for urban masterplans.

The urban block is flexible physical element that take on many physical forms, shapes, scales, and densities. Indeed, Tarbatt and Street-Tarbatt (2020: 71) describe how: *"It is the potential for adaptation and innovation of the block that confirms both the continued relevance of the perimeter block to urban designers."* Careful planning and design of an urban block can create coherence and character to the streets, squares, and other public spaces with prime examples being the European cities of Barcelona and Paris. The centre of the block can also provide important private spaces for gardens, play areas, and other communal activities within urban environments and thereby providing a clear articulation and legibility between the public and private realms. Significantly, the urban block is also able to accommodate a wide variety of synergistic uses and activities such as town houses or apartments, shops, offices, public functions, leisure facilities, and healthcare centres (see the section on Mixed Use and Multi-Dimensional). Importantly, the urban block structure also allows for transformation over time whilst also enabling variety in the form, scale, and aesthetic style of its architecture, thereby enabling a diverse and interesting townscape.

The ideal scale of urban blocks is a much-discussed topic. Jane Jacobs (1961) argues for the importance of frequent streets, and small-scale urban blocks in order to enable greater permeability and choice of routes through an urban environment. In contrast, Llewelyn Davies in the *Urban Design Compendium* (2000) recommends large perimeter blocks of 90 metres by 90 metres to provide a good trade-off between biodiversity and other considerations. There are strong arguments for both small and larger urban blocks; however, it is important for urban designers to carefully consider the morphological context of any block and to be informed by neighbouring blocks and street patterns within the respective city. Jacobs (1961) also emphasises the importance of active street corners and suggests that the multiplicity of routes created by smaller urban blocks provides choice and variation but also helps to facilitate small commercial opportunities on street corners. As key physical points within the built environment where there is choice regarding direction of movement, street corners can be important places that make towns and cities more legible. Also, the location of uses such as shops, cafes, or bars on street corners can create important landmarks which not only help people to orientate themselves within a neighbourhood but also act as places where people arrange to meet, or for chance encounters.

FURTHER READINGS

Bürklin, T. & Peterek, M. (2007) *Basics Urban Building Blocks*, Birkhäuser: Basel.

Llewelyn-Davies (2000) *Urban Design Compendium*, with English Partnerships and The Housing Association: London.

Palsson, K. (2023) *Urban Block Cities: 10 Design Principles for Contemporary Planning*, DOM: Freiburg.

Rowe, P.G., van den Berg, H.J. & Wang, L. (2019) *Urban Blocks: History, Technical Features, and Outcomes*, Scholars' Press: Beau-Bassin.

Tarbatt, J. & Street-Tarbatt, C. (2020) *The Urban Block: A Guide for Urban Designers, Architects and Town Planners*, RIBA Publishing: London.

THE GLOBAL CONTEXT

One of the most important considerations of urban design projects is the many contexts that make all locations and therefore projects

Figure 4.13 Context is a primary consideration in urban design and the public square as a formal marketplace in common within many cultures (Old Market Square, Nottingham, UK).

Source: Photograph taken by the authors.

unique. Whilst the significance of these contexts will be discussed later (see also the section on Contextually Responsive), it is important to emphasise that across the world, there are different cultural responses to peoples' needs and requirements that significantly impact not just the physical appearance of cities and their public realms, but also the way that local communities use public space. Indeed, although peoples' basic needs around the world, somewhere to live, work, socialise, shop, study, etc., are similar, there are unique urban solutions to how these are organised and shaped that reflect different contexts such as the social, cultural, political, religious, economic, and climatic conditions (Figures 4.13 and 4.14).

These contexts manifest themselves in the design, location, scale, and purpose of the common physical urban design elements such as the street and the square. Significantly, Carmona (2021: 56) stresses how *"it is increasingly important to respect the cultural diversity that continues to exist because this permits authentic local distinctiveness."* In a Western context for example,

Figure 4.14 Public spaces can also accommodate more informal markets such as night markets (Garden Night Market, Tainan City, Taiwan).

Source: Photograph taken by the authors.

the public square has throughout history held critical and central prominence in cities as a focal point for various different activities and purposes. In Europe, for example, the Rynek Główny (Main Market Square) in Kraków, the Piazza del Campo in Sienna, and the Old Market Square in Nottingham are typical of the importance and centrality of the traditional marketplaces as places for trade and commerce. Other well-known public squares such as the Piazza San Pietro in Rome represent the global centre of the Catholic religion, whilst the Piazza della Signoria in Florence celebrates the power of art and politics. In other cities, major public squares were planned as multi-purpose spaces, such as the Plaza Mayor in Madrid, or Rittenhouse Square in Philadelphia. In other parts of the world, such as Latin or South America, the major public squares in cities, such as Havana, Lima, and Santiago, were often called the Plaza de Armas and were created as places to demonstrate military strength and refuge. In other cultures, the public square as a representation of political power has resulted in larger spaces, such as Heroes' Square in Budapest, or Tiananmen Square in Beijing.

Public squares can also be found in Middle Eastern and South-East Asian cities, often but not always influenced by past periods of colonisation, but the focus of urban life, primarily due to cultural and climatic contexts, tends to be on the streets and covered public places. Nevertheless, the public square has still played important roles in the evolution and urban form of many cities in these regions. Notable examples include Naqsh-e-Jahan Square in Isfahan, Tahrir Square in Cairo, or Dusit Palace Plaza in Bangkok, Merdeka Square in Jakarta, and Dataran Merdeka in Kuala Lumpur. Nevertheless, in these cultures, the streets and the covered marketplaces, such as souks and bazaars, were more representative of the places for trading and social interaction. As such, there are many excellent examples of diversity of street markets as unique examples of the public realm such as Itaewon Market in Seoul, Chatuchak Market in Bangkok, The Raohe Night Market in Taipei, the Luang Prabang Night Market in Laos, specialist markets such as the Tsukiji Fish Market in Tokyo, or markets in unique locations such as the Train Market in Bangkok, and the floating markets in Cai Rang in Vietnam, and Damnoen Saduak in Bangkok. Interestingly, whilst Europe also has its famous covered market halls, such as La Boqueria in Barcelona, Mercato Centrale in Florence, and Mercado da Ribeira in Lisbon, the great souks and bazaars of the Middle East and North Africa are at another level in terms of public spaces with dramatic scale, colour, and activity. These are epitomised by many famous examples, such as the Grand Bazaar in Istanbul, Jemaa El-Fnaa Bazaar in Marrakech, and Souq Waqif in Doha (Figures 4.15 and 4.16).

Understanding and responding to the unique contexts of urban environments is therefore imperative in the urban design process. Indeed, it is these particular regional and global contextual differences that help to distinguish different places and make the public realm a unique celebration of the evolution of different cultures and communities upon cities. This is particularly important in an era when large design companies work in an international market (Loew 2012). Indeed, in an era when concerns are increasingly expressed about the impact of globalisation and standardisation upon the built environment, it is especially important that the design and regeneration of the public realm reflects the communal

Figure 4.15 In some contexts, urban squares can have more of a ceremonial role as a celebration of power or as a place of display (Tiananmen Square, Beijing, China).

Source: Photograph taken by the authors.

Figure 4.16 Markets are great displays of local culture and in some countries the grand market hall is a dominant urban feature (Central Market, Valencia, Spain).

Source: Photograph taken by the authors.

and cultural values together with the societal structures that are unique to their specific locations.

FURTHER READINGS

Chen, F. & Thwaites, K. (2016) *Chinese Urban Design: The Typomorphological Approach*, Routledge: London.

Cookson Smith, P. (2023) *Writings on the Asian City*, ORO Editions: Hong Kong.

Gharipour, M. (ed.) (2016) *Contemporary Urban Landscapes of the Middle East*, Routledge: London.

Kiang, H.C., Liang, L.B. & Limin, H. (eds.) (2010) *On Asian Streets and Public Space*, NUS Press: Singapore.

Loew, S. (ed.) (2012) *Urban Design in Practice: An International Review*, RIBA Publishing: London.

Wiedmann, F. & Salama, A.M. (2019) *Building Migrant Cities in the Gulf: Urban Transformations in the Middle East*, I.B. Taurus: London.

PRINCIPLES OF URBAN DESIGN

Public spaces are complex, organic things and design is only a small fraction of what goes into making a great public space. Indeed, the on-going use of the public realm together with its management and maintenance all influence the success of the streets and squares of an urban environment. The breadth of the urban design process and the wide spectrum of actors engaged in the public realm therefore mean that ensuring the creation, refurbishment, and curatorship of public spaces is a complex and multi-faceted process.

In enabling the creation of successful public spaces, urban designers need to consider a series of fundamental principles that provide the foundation for its theory and practice. These principles are generally universal, and they can be applied, subject to appropriate contextualisation, to a diverse range of typologies and locations. Indeed, with sensitive adjustment for cultural, environmental, and climatic conditions, they are applicable to different places and communities around the world. It is important to acknowledge, however, that all urban design solutions, whilst applying these principles, should be site- and context-specific if we are to create unique, resilient, and successful places that will foster positive and enjoyable experiences for all users of the public realm.

Importantly, it should also be emphasised that not all successful public spaces will display the successful application of all of these principles; however, it is highly likely that they will demonstrate a significant number of them. Although many of these fundamental urban design principles are intrinsically connected, for ease of understanding, in this chapter they will be organised into four sections: physical principles, functional principles, social principles, and operational principles. These principles will be explained by focusing specifically on the main components of the public realm: the street, the public square, and the urban block.

FURTHER READINGS

Carmona, M. (2021) *Public Places Urban Spaces: The Dimensions of Urban Design*, Third Edition, Routledge: New York.

Cowan, R. (2021) *Essential Urban Design: A Handbook for Architects, Designers and Planners,* RIBA Publishing: London.

Sim, D. (2019) *Soft City: Building Density for Everyday Life*, Island Press: Washington.

PHYSICAL PRINCIPLES OF URBAN DESIGN

From a physical design perspective, it is commonly recognised that in most contexts, the considered application of the basic urban design components of streets, squares, and urban blocks will help to ensure a successful public realm and urban form. There are, however, a number of key principles that urban designers should consider during the design process to help ensure that both new and refurbished urban places meet the needs and expectations of the various users of the public realm. Whilst in practice these are intertwined, they are organised here into six themes for ease of explanation: contextual responsiveness; density; active edges; connectivity and legibility; flexibility and adaptability; and incremental change and growth.

CONTEXTUAL RESPONSIVENESS

Urban design projects, whether they are focused on improving the existing public realm and townscape or the creation of new places, will have existing contexts to which they should respond sensitively.

Indeed, the public realms of towns and cities around the world are influenced by an extensive array of contexts that will influence not just their appearance and the way that they are used but also the way that they are experienced and what these places mean to locals and visitors. These will vary from place to place, city to city, region to region and include at least some of the following contexts: social, economic, cultural, physical, environmental, climatic, time, etc. All urban design projects should carefully consider these contexts in achieving the fundamental goal of emphasising and strengthening the unique characteristics of a place. Doing so, can enhance the aesthetic appeal and historical context which in turn can help to create a sense of identity and place, emphasise continuity, and foster community belonging. To achieve this successfully, it is imperative that urban designers recognise and value the differences and uniqueness between places to ensure that projects protect and enhance local and regional attributes.

Whilst contemporary urban design should not seek to replicate the past, important lessons and precedents can be drawn from how places and contexts have evolved through history. Indeed, Black et al. (2024: 10) argue *"that urban design is needed now more than ever, to ensure places remain contextual, responsive and empowering for people."* As such, it is imperative for designers to respect, respond to, and be informed by the heritage of a particular place in determining a successful design solution. Whilst the needs and demands placed upon the public realm in the past may have changed beyond recognition, it remains important to apply lessons from the past to the conditions and requirements of the present. Indeed, projects that embrace a culturally responsive approach will integrate physical, social, and cultural heritage to enable public spaces to facilitate vibrant contemporary urban experiences whilst celebrating their historical significance.

FURTHER READINGS

Black, P. Martin, M., Phillips, R. & Sonbli, T. (2024) *Applied Urban Design: A Contextually Responsive Approach*, Routledge: New York.

Carmona, M. (2021) *Public Places Urban Spaces: The Dimensions of Urban Design*, Third Edition, Routledge: New York.

Gehl, J. (1996) *Public Spaces – Public Life*, Island Press: Washington DC.

DENSITY

Urban density typically refers to how intensively land is developed and can be expressed in terms of population or developed floor area per area. Density varies considerably between settlements, regions, and countries; however, urban areas account for the majority of population growth and cities are becoming denser. Urban design, therefore, has a central role to play in ensuring that this is accommodated through the appropriate design of the public realm but how this engages with existing and new housing, employment, transportation, and other services. Indeed, Jane Jacobs (1961) was a strong advocate of medium- to high-density development in order to ensure a sufficient concentration of people to help maintain vitality (see Chapter 3). CPRE (2019: 2) also advocates that: *"Bigger concentrations of people stimulate and support the provision of more services and facilities."*

The effective management of density is central to the creation of compact, well-planned cities, and densification is perceived as a fundamental strategy for creating sustainable accessibility. Indeed, many cities across the world are now actively promoting policies towards more compact urban development (RICS 2021). This has largely been a reaction to the propensity towards lower-density development, suburbanisation, land-use zoning, urban sprawl, and reliance on the motor car, experienced in the mid-to-late twentieth century. Significantly, more compact cities (see Chapter 2) aim to have a good mix of uses and activities to enable as many people as possible to have easy access to services either by foot or by public transport. Indeed, a critical mass of people enables urban areas to be better connected by efficient and affordable public transport systems. This in turn helps to make a city more inclusive in terms of access to local housing, education, employment, and services.

Despite historically being associated with overcrowded urban metropolises, well-designed medium- to high-density urban environments are now accepted as sustainable solutions for cities around the world. Significantly, the Essex Design Guide (Essex County Council 2021) argues that: *"High density does not mean high-rise towers. Sustainable neighbourhoods use density to create intensity of public space, enlivening the streetscape by creating desirable spaces that engender strong communities."* Indeed, having more people living, working, studying, and taking part in recreational activities helps to make places safer, justifies more services, helps to make uses and activities more economically viable, promotes walkability and public transportation, and reduces car dependency (see also the sections on Walkability, Multi-Modal Transit, and Economic Value).

FURTHER READINGS

CABE (2005) *Better Neighbourhoods: Making Higher Densities Work*, Commission for Architecture & the Built Environment: London.

Greater London Authority (2016) *London Plan Density Research: Lessons from Higher Density*, GLA: London, www.london.gov.uk/sites/defa ult/files/project_2_3_lessons_from_higher_density_development.pdf (accessed 25 August 2023).

RICS (2021) *Urban Density: Promoting sustainable development*, Royal Institute of Chartered Surveyors, www.rics.org/news-insights/wbef/ urban-density-promoting-sustainable-development-part-1 (accessed 18 November 2024).

Sim, D. (2019) *Soft City: Building Density for Everyday Life*, Island Press: Washington DC.

Urban Land Institute & Centre for Liveable Cities (2013) *10 Principles for Liveable High Density Cities. Lessons from Singapore*, Centre for Liveable Cities: Singapore.

ACTIVE EDGES

The buildings that frame all public places, including streets and squares, should contain as many active facades as possible. These active edges or active frontages are created where the functions or activities inside the buildings have physical and visual interaction with the public realm. This involves a transparency whereby people actively move in and out of the buildings together with clear lines of view between the users and activities occurring internally and those occupying the public space. This is especially important at the ground floor level, where having functions and activities that engage with the space not only activates the space, thereby

Figure 4.17 Streets with inactive facades providing no natural surveillance create an unattractive and unwelcoming environment that often convey a perception of being unsafe.

Source: Photograph taken by the authors.

making it more attractive and interesting but also makes it feel safer due to the resultant natural surveillance (see also the section on Safety and Natural Surveillance). Indeed, having ground floor uses that involve regular movement in and out of the buildings creates physical activity that also provides visual interest in terms of what is occurring within the building and helps to create a sense of vibrancy within the public realm.

Figure 4.18 Streets that have active edges with high levels of natural surveillance tend to be more attractive and feel safe (Huangshan, Anhui Province, China).

Source: Photograph taken by the authors.

In addition, land uses and activities that can 'spill-out' or extend into the public realm, such as seating and tables for bars, cafes, and restaurants, or stalls and displays outside of shops can further stimulate life and visual diversity for the users of public spaces. To enhance this further, frequent and active building entrances will further stimulate pedestrian movement. In contrast, blank facades without active or engaging functions on the ground level will not enliven the street, provide any natural surveillance, or encourage people to walk alongside them. It is therefore fundamental for urban

Figure 4.19 Ground floor functions that spill-out into the space help to enliven streets, provide opportunities for people watching, and enhance activity (Rue Soufflot, Paris, France).

Source: Photograph taken by the authors.

designers to stimulate life and activity in the public realm by carefully considering the impact of ground floor and 'spill-out' activities on the urban experience and ensure that the designs of the buildings and public spaces are mutually supportive (Figures 4.17–4.20).

The active edge and frontage concept for the public realm can be further enhanced by utilising the inner and outer space concept to the design of streets and squares. Indeed, the edge of the public spaces can be further strengthened by having a complementary inner or neighbouring space to further enhance the user experience. For example, where there might be coffee outlets, food takeaway vendors, or bookshops on the edge of the space, there can be an adjacent space with seating, and shelter for people to drink their coffee, eat their food, or read their books. Achieving this can create a curated and vibrant urban scene that will be inviting and attract further users resulting in an urban environment with a critical mass of people that provides activity and safety to the space whilst also helping the economic viability of the local traders.

Figure 4.20 Shops and cafes spilling out onto the street provide both visual interest and increase natural surveillance (Ruga dei Oresi, Venice, Italy).

Source: Photograph taken by the authors.

FURTHER READINGS

Jacobs, J. (1961) *The Death and Life of Great American Cities*, Random House: New York.

Llewelyn-Davies (2000) *Urban Design Compendium, with English Partnerships and* The Housing Association: London.

Sim, D. (2019) *Soft City: Building Density for Everyday Life*, Island Press: Washington.

CONNECTIVITY AND LEGIBILITY

All successful public spaces, both streets and squares, need to be well connected and easy for people to find and discover, whether though intention or by chance discovery. It is therefore important that the overall network of routes and circulation provide easy access, choice, and safety for people moving to and through an urban environment. Indeed, for public places to be successful they need to be easy to find and therefore designing the public realm to assist people in finding their way around needs to be fundamental to the urban design process. Whilst many public spaces, such as squares, are destinations in their own right, others are discovered as people move through the urban environment. These other spaces, streets and squares, therefore need to be linked into a legible network if they are to be well-used and therefore successful elements of a city's fabric (Figures 4.21 and 4.22).

To ensure a legible townscape, a clear hierarchy of streets and public spaces needs to be established to enable pedestrians and cyclists to move conveniently, enjoyably, and safely around the city. This can be achieved through the creation of a well-designed layout but also further enhanced through wayfinding signage to improve accessibility, orientation, and the connectivity of spaces and activities. It is also important that all public spaces are easy to access, particularly by pedestrians but also by public transport, and therefore, it is important to remove any barriers to access such as wide and busy roads (see also the sections on Walkability, and Multi-Modal Transit) (Project for Public Spaces 2022). Having regular and safe pedestrian and cycle crossing points, well-located public transit stops, and slowing down the speed of any vehicular traffic are all measures that can ensure people can not only locate but access a city's public spaces.

FURTHER READINGS

Llewelyn-Davies (2000) *Urban Design Compendium, with English Partnerships and* The Housing Association: London.

Figure 4.21 Providing vistas of well-known landmarks can enhance legibility and wayfinding within cities (Calle Padre Hortas Cáliz, Jerez de la Frontera, Spain).

Source: Photograph taken by the authors.

Matan, A. & Newman, P. (2017) *People Cities: The Life and Legacy of Jan Gehl*, Island Press: Washington DC.

Sim, D. (2019) *Soft City: Building Density for Everyday Life*, Island Press: Washington.

FLEXIBILITY AND ADAPTABILITY

The long-term evolution of towns and cities highlights the importance of designing public realms that have characteristics that facilitate change. Indeed, whilst, in most cities, buildings become

Figure 4.22 Landmarks that terminate vistas help to induce movement through the urban environment (Yasaka Pagoda, Kyoto, Japan).

Source: Photograph taken by the authors.

obsolete and are replaced over time, public spaces, including streets, squares, and parks, tend to last for a considerably longer time. As such, the most successful public spaces tend to be flexible and adaptable to meet existing needs and demands whilst also accommodating future changes and expectations. Indeed, the design of the public realm should be flexible to accommodate a variety of daily activities together with other occasional events. In

Figure 4.23 Successful public spaces are inherently flexible and able to host a wide range of different events and activities (Place de l'Hôtel de Ville, Paris, France).

Source: Photograph taken by the authors.

different places, these might happen and change on a daily, weekly, seasonal, or annual basis and spaces need to have an inherent flexibility to enable them to successfully occur (see also the section on Mixed-use and Multi-Dimensional). The design of flexible spaces also needs to enable infrastructure and services, such as electricity, water, and drainage, to be available for various potential pop-up activities to take place. To ensure the long-term success and viability of public spaces, it is important that they are adaptable to allow for the future enhancement of the space and are therefore flexible and able to embrace and withstand the changes that will arise over time (Figures 4.23–4.25).

FURTHER READINGS

Bishop, P. & Williams, L. (2012) *The Temporary City*, Routledge: London.
Madanipour, A. (2017) *Cities in Time: Temporary Urbanism and the Future of the City*, Bloomsbury: London.
Wunderlich, F.M. (2023) *Temporal Urban Design: Temporality, Rhythm and Place*, Routledge: London.

Figure 4.24 New public spaces can be created through the adaptive re-use of brownfield sites that previously had industrial uses (Coal Drops Yard, London, UK).

Source: Photograph taken by the authors.

INCREMENTAL CHANGE AND GROWTH

Historically, most towns and cities expanded and changed incrementally at a relatively slow rate often over many centuries. Alongside physical changes to buildings and infrastructure, communities also tended to change at a slow rate leading to social cohesion and strong place attachment. This enabled people to actively engage in changes to their neighbourhood and helped to give them a sense of 'ownership' and 'responsibility' for the place. More rapid and wholescale developing since the mid-to-late twentieth century has led to many advocating a more considered and sensitive approach to urban change and growth. Indeed, Christopher Alexander et al. (1977: 3) argued for the benefits of place-making through incremental change and suggested that through piece-meal growth, places *"...can emerge gradually and organically, almost of their own accord, if every act of building large or small takes on the responsibility for gradually shaping its small corner of the world."*

COMPONENTS AND PRINCIPLES OF URBAN DESIGN 101

Figure 4.25 Abandoned transport infrastructure in cities, such as former railway lines, can also be adapted into successful public places (The High Line, New York, USA).

Source: Photograph taken by the authors.

Incremental urban development is considered to be more socially, economically, and environmentally sustainable through the promotion of gradual changes to urban areas. This can be achieved by facilitating increases in building size, intensity of activities, and the refurbishment or replacement of buildings in a more incremental manner so that change is less drastic, and the overall character of an area remains intact. Additionally, making use of and adapting existing infrastructure can help with the sustainability challenges facing cities. The smaller-scale and gradual evolution of the built environment, such as infill projects and adaptive re-use, rather than wholescale redevelopment therefore has many benefits. This can also enable community retention and create more opportunities for all sectors of society in terms of access to and affordability of housing, a range of employment opportunities, and access to public services (see also Varied Tenure and Condition).

FURTHER READINGS

Bishop, P. & Williams, L. (2012) *The Temporary City*, Routledge: London.
Jacobs, J. (1961) *The Death and Life of Great American Cities*, Random House: New York.
Medrano, L., Recamán, L. & Avermaete, T. (eds.) (2021) *The New Urban Condition: Criticism and Theory from Architecture and Urbanism*, Routledge: New York.
Tiesdell, S., Oc, T. and Heath, T. (1996) *Revitalizing Historic Urban Quarters*, Architectural Press: Oxford.
urbanNext (2025) *Flexible Urbanisms: Towards an Incremental Urbanism*. https://urbannext.net/flexible-urbanisms/

FUNCTIONAL PRINCIPLES OF URBAN DESIGN

When the functional characteristics, in terms of the uses and activities with an urban environment, are varied and complementary, then the public spaces are much more likely to become attractive, healthy, and safe destinations and places in which people will choose to gather, interact, and socialise. To achieve this, again there are key principles that urban designers should consider to help ensure sustainable and resilient public realms. These are organised here into four themes related to the use and movement of people: mixed-use and multi-dimensional; varied tenure and condition; walkability; and multi-modal transit.

MIXED-USE AND MULTI-DIMENSIONAL

Historically, urban environments were typically organised out of necessity as mixed-use areas with the buildings at street level being devoted to business and trade, whilst the upper floors were used for ancillary activities or residential use (Roberts & Lloyd-Jones 1997). Significantly, since the late twentieth century, mixed-use development has again become fundamental to the organisation and planning of urban areas as a reaction to the mono-functional zoning of land uses prevalent in earlier modernist planning visions (see Chapter 2). In the contemporary city, mixed-use development involves the sensitive combination of residential, employment, education, leisure, and public activities into a building, urban block, street, or neighbourhood. The organisation and relationship of these different functions can be achieved in many ways, including the vertical stacking of various land uses, or co-locating different functions as neighbours along a street as in the traditional high street. Significantly, a mix of land uses and activities is not imperative in every building or urban block. It is important, however, that there is an adequate mix along a street or within a neighbourhood to create vitality and intensity in the public realm through use by a wide range of people. In addition to people living and working there, others should be attracted to other opportunities such as shops, restaurants, cafes, bars, hotels, gyms, schools, and civic functions.

> *Public space is inherently multi-dimensional. Successful and genuine public spaces are used by many different people for many different purposes at many different times of the day and the year. Because public spaces harbor so many uses and users – or fail to do so – they are also where a staggering cross-section of local and global issues converge.* (Project for Public Spaces 2022: 1)

One of the key criteria for mixed-use development is the compatibility and synergy of the different functions and activities. Historically, industrial uses that caused noise and pollution were often located adjacent to a catchment workforce; however, in contemporary cities, such factories rarely exist, and the majority of land uses can be successfully co-located. Indeed, residential, employment, education, retail, tourism, cultural, and civic functions can all have social and economic synergies and in turn promote the wider benefits of mixed-use neighbourhoods. The importance of diversity is emphasised by the Urban Land Institute (2019: 1) who state that mixed-use development should be characterised by *"three or more significant revenue-producing uses (such as retail/entertainment, office, residential, hotel, and/or civic/cultural/recreation)."* The Project for Public Spaces (2023d) develops this further with their *Power of 10+* concept which articulates how having a mix of land uses and a variety of activities within close proximity has multiple benefits in the creation of successful places (see also the section on Project for Public Spaces in Chapter 3). Seasonal strategies can also help to attract a more diverse range of users at different times of the year. Such an approach can help to facilitate change based on the season and give users ever-changing experiences during the year. In the summer, for example, a public space may have an urban beach or outdoor film shows, whilst in the winter, a square may host an outdoor skating rink, or a festival market.

An increasingly popular approach for animating the public realm with short-term or temporary activities. Indeed, the past 20 years or so has seen significant interest in the concept of temporary urbanism with city authorities, organisations, and local communities organising and hosting a wide range of pop-up events such as markets, food markets, entertainment, shops, exhibitions, ice skating, beaches, fairs, pop concerts, sporting events, film shows, gardens, and libraries. If at least some, public spaces are designed to have inherent flexibility, then one day a space can, for example, host a farmer's market, the next day have a pop-up netball tournament with a concert in the evening, and the following week hold a community festival, or an open-air market. Such a strategy can result in an ever-changing visitor experience and one that can attract different users to the public realm at different times. These short-term activities, which can be managed (organised), unmanaged (unofficial), or spontaneous, can provide important animation to public spaces that attracts both existing and new users to use and activate the city. Another increasingly popular approach to temporary urbanism is to organise events as affordable pilot projects to test their success within public spaces. These smaller-scale urban design projects can be implemented for short time periods and, often using low-cost materials, to help inform public decision-making, and allow the public to experience and test the project prior to the implementation of a potentially more expensive longer-term project (Figures 4.26 and 4.27).

Cities have traditionally evolved because of many foreseen and unforeseen circumstances such as social changes, technological advancements, economic situations, and changes in taste or fashion. A significant benefit of mixed-use development is its ability to enable such changes over time in a sustainable manner. Coupland (1997: 1) also identifies the important role of mixed-use

Figure 4.26 Mixed-use developments such as shophouses create many different reasons for people to visit a place and help to stimulate activity throughout the day and night (Smith Street in Singapore).

Source: Photograph taken by the authors.

Figure 4.27 Neighbourhoods that contain many different functions tend to attract a wide variety of different users and encourage people to walk from one activity to the next (Dafen Village, Shenzhen, China).

Source: Photograph taken by the authors.

development in creating sustainable cities, suggesting *"...that by increasing the mix of land uses, and especially residential uses, residents will lead more sustainable lifestyles, using their cars less."* More specifically, the benefits include: (i) creating more reasons for people to need or want to be in that place; (ii) attracting a broader range of users; (iii) making the place more economically viable and sustainable; (iv) making the area safer; (v) encouraging walking or cycling whilst reducing reliance on car travel; and (vi) creating more visual interest. Each of these benefits has in turn many other positive spin-off impacts such as: reducing traffic congestion; reductions in energy consumptions; health benefits; and inclusivity for all residents that are important considerations for urban design (see also the sections on Inclusive and Accessible, Healthy and Environmentally Responsive, and Economic Value). Also, from a sustainability and continuity perspective, where cities have had large areas of mono-functional development, such as factory sites, industrial docklands, energy plants, and shopping malls, and this land-use has subsequently become obsolete it has tended to leave large swathes of land and buildings abandoned and unused. Trying to find new functions of a similar scale is difficult and finding sufficient smaller functions to occupy the left-over sites is also challenging with few willing or able to be pioneers locating in the middle of an abandoned part of the city. However, with mixed-use development, when one function becomes obsolete or moves out it is considerably easier to find a replacement as the majority of the area will still be active and successful and also therefore more sustainable.

Building upon the benefits of mixed-use development, the 15-minute city concept (see the section on the 15-Minute City in Chapter 2) advocates walkable pedestrian-centred neighbourhoods in which most basic needs and services, such as healthcare, shopping, education, work, and leisure are accessible within a 15-minute walk or bicycle ride from any point in the city. A large part of the idea centres around human-centred design and the creation of more sustainable, healthy, and inclusive cities. Despite the homogeneity of the concept, applying it to different existing places, neighbourhoods, densities, and cultures should result in contextually sensitive solutions that can reflect specific local situations and peculiarities and thereby also help to sustain or even revive local distinctiveness and character.

FURTHER READINGS

Carmona, M. & Wunderlich, F.M. (2012) *Capital Spaces: the Multiple Complex Public Spaces of a Global City*, Routledge: London.

Llewelyn Davies (2000) *Urban Design Compendium*, English Partnerships and the Housing Corporation: London.

Project for Public Spaces (2008) *Our Approach to Mixed-Use*, December. www.pps.org/article/mixeduseapproach (accessed 27th July 2023).

Urban Land Institute (2019) *Understanding Mixed Use and Multi Use*. https://knowledge.uli.org/-/media/files/reading-list/reading-list-pdfs/readinglist_mixeduse_vl.pdf?rev=cf67335ad1b44772b6f20bb074043a15 (accessed 23rd June 2023).

VARIED TENURE AND CONDITION

Most towns and cities have evolved gradually over hundreds of years, and as a result, they typically possess buildings and spaces that display different characters, styles, and ages. In addition, there will be buildings, that contain a range of uses and activities, in varying states of repair and condition that in turn will command different sale or rental values. In terms of social and economic sustainability, having properties in a neighbourhood with a varying level of affordability is important in order to create opportunities and inclusivity for the local communities. From a residential perspective, having homes of varying sizes, condition, and affordability enables poorer and more affluent residents to live in a mixed community, which in turn helps to create more diverse, cohesive, and interesting communities. Indeed, the Ministry of Housing, Communities & Local Government (2021: 35) identifies how *"well-designed neighbourhoods provide a variety and choice of home to suit all needs and ages. This includes people who require affordable housing or other rental homes, families, extended families, older people, students, and people with physical disabilities or mental health needs."*

From the perspective of the local economy and employment opportunities, having a range of business premises, in terms of size and cost, can help to provide important places for small- to medium-scale enterprises, and start-up businesses (Jacobs 1961). Such properties can also facilitate pop-up business initiatives which can be a catalyst for more permanent ventures and also help to create economic and social vitality within a neighbourhood. Oldenburg (1989), in his book *The Great Good Place*, discusses how many of these small business ventures become essential places for local people to gather and interact and thereby contribute

Figure 4.28 Neighbourhoods with buildings of different size, condition, and tenure help to promote opportunities for small local businesses which can boost the local economy and also create places with unique characteristics (Tianzifeng, Shanghai, China).

Source: Photograph taken by the authors.

significantly to community building and neighbourhood vitality (Figures 4.28).

In addition to the economic and social benefits, of having buildings of varied condition and tenure, older buildings can help to contribute to the character of a place and also represent its historical and cultural heritage. As such, the diversity of buildings can help to define the qualities that make a place different from other areas in the city and contribute to neighbourhood's place identity. This can be maintained and strengthened through conservation approaches and also through the adaptive re-use, or upcycling, of redundant buildings. Indeed, many buildings become obsolete over time in terms of their original functions yet through careful adaptation and repurposing they can be transformed for new uses or functions (Tiesdell et al. 1996). When done sensitively, this can provide some townscape continuity in terms of the physical building by celebrating its historical or cultural qualities. In addition, the retrofit of existing buildings can be far more sustainable solution and help to meet net-zero carbon targets. The significance of this approach is emphasised by the fact that *"it is estimated that 80 per cent of buildings currently standing will still be in use in 2050"* (UKGBC n.d.). The upgrading of the environmental performance and the containment of the embodied energy within these original buildings as part of the adaptive re-use process can therefore also play an important role in environmental sustainability.

FURTHER READINGS

ARUP (2020) *Transform and Reuse: Low-Carbon Futures for Existing Buildings*, ARUP: London.

Baker-Brown, D. (2024) *The Re-Use Atlas: A Designer's Guide Towards a Circular Economy*, RIBA Publishing: London.

Jacobs, J. (1961) *The Death and Life of Great American Cities*, Random House: New York.

Jacobs, J. (1970) *The Economy of Cities*, Vintage Books: New York.

Oldenburg, R. (1989) *The Great Good Place: Cafés, Coffee Shops, Community Centers, Beauty Parlors, General Stores, Bars, Hangouts, and how They Get You Through the Day*, Paragon House: New York.

Tiesdell, S., Oc, T. and Heath, T. (1996) *Revitalizing Historic Urban Quarters*, Architectural Press: Oxford.

WALKABILITY

Urban designers should focus primarily on designing cities for people not cars to create human-centric public space. Whilst private vehicles are arguably a necessity in most cities, more attractive and successful cities prioritise the walking, cycling, and public transport use. If designed well, local car movement can be integrated into the public realm; however, the priority should be given to pedestrians and cyclists. The creation of more walkable cities is based on many of the same principles and has close synergies with the 15-minute city concept (see the section on the 15-Minute City in Chapter 2). Walkable cities are pedestrian-friendly environments that encourage people to forego the convenience of car travel in order to travel on foot (Figures 4.29–4.31).

The benefits of making cities more walkable, have been widely discussed and extensively researched (Transport for London 2010; ARUP 2016), include reducing carbon emissions and pollution, reducing traffic congestion, encouraging a healthier and fitter population, improving the viability of small shops and businesses, increased inclusivity and opportunities, etc. Indeed, cities will become far more environmentally, economically, and socially sustainable if people make a positive decision to walk. Importantly, increased pedestrian footfall also results in more passing trade that can help to make shops, cafes, and other local businesses more economically viable and these in turn will attract more people to frequent the streets and squares on which they are located (Forsyth 2015). A spin-off benefit of this is that the area will also become safer as more people on foot means more natural surveillance. Other important outcomes of more people choosing to walk are the health benefits to society. Indeed, governments could save considerable funding every year in, for example, obesity-related healthcare costs if people could be persuaded to walk more each week. Achieving improved walking environments in cities does, however, require some additional costs for the widening of footpaths, provision of seating, shelter, and public toilets, however, this far less expensive than constructing and maintaining highway infrastructure.

Significantly, unlike car travel, walking is also an accessible form of transportation, for all but a minority of people with a disability that prevents it, that does not discriminate between various socio-economic groups. It is also important that walking routes efficiently connect key destinations such as schools, shops, places of employment, and leisure activities. Encouraging people to walk more will also require better signage to improve wayfinding and enable people to conveniently and safely navigate through urban environments. Obviously, as discussed in the next section on

Figure 4.29 Poor quality pavements especially with poorly located street furniture, lampposts, and signs can discourage walking and cycling and be very difficult for people that are less mobile (Kuala Lumpur, Malaysia).

Source: Photograph taken by the authors.

multi-modal transit, it is imperative to achieve a workable balance in terms of how people move within cities. Indeed, from a walkability perspective expected distances of travel need to be realistic and therefore well-integrated with other modes. If achieved successfully, however, cities can become healthier, safer, and more inclusive environments.

COMPONENTS AND PRINCIPLES OF URBAN DESIGN 113

Figure 4.30 The elderly, children, and those with physical challenges can find urban environments hostile and dangerous if they are not designed appropriately.

Source: Photograph taken by the authors.

FURTHER READINGS

Appleyard, D., Gerson, S. & Lintell, M. (1981) *Livable Streets*, University of California Press: Berkeley.

ARUP (2016) *Cities Alive: Towards a Walking World*, 27 April, ARUP: London.

ARUP (no date) *Delivering Sustainably Walkable Neighbourhoods: An Evidence-led Approach*. www.arup.com/insights/delivering-sustainably-walkable-neighbourhoods (accessed 20 September 2024).

ARUP (no date) *Walkable Cities are Better Cities*. www.arup.com/insights/delivering-sustainably-walkable-neighbourhoods (accessed 23 June 2023).

Figure 4.31 Successful walkable urban environments tend to be well-maintained, safe, and welcoming places (Plaza de la Yerba, Jerez de la Frontera, Spain).

Source: photograph taken by the authors.

Bruntlett, M. & Bruntlett, C. (2021) *Curbing Traffic: The Human Case for Fewer Cars in Our Lives*, Island Press: Washington DC.

Gehl, J. (2011) *Life Between Buildings: Using Public Space*, Island Press: Washington DC.

Speck, J. (2013) *Walkable City: How Downtown Can Save America One Step at a Time*, North Point Press: New York.

Speck, J. (2019) *Walkable City Rules: 101 Steps to Making Better Places*, Island Press: Washington DC.

MULTI-MODAL TRANSIT

Whilst facilitating walkability is a priority for urban designers, this needs to be part of a holistic approach to the convenient, efficient, and equitable movement of people through cities. In doing so, urban design should focus on prioritising active and sustainable modes of transport within a range of mobility options. As such, choice is the important consideration and the most effective and sustainable solutions are those were people make a positive choice to walk, cycle, or take public transport in their daily lives. This will only be achieved when these options are comfortable, affordable, reliable, and safe. To achieve this requires suitably designed high-quality public realms, in particular street networks, and appropriately located land uses and activities (see also the section on Mixed-Use and Multi-Dimensional).

To achieve this, streets must be designed to serve different modes and provide multiple mobility options for its users. Multi-modal streets help to make cities more efficient if they provide convenient accessibility both within neighbourhoods and are well-connected to a city-wide transit and cycling network. When successfully designed, multi-modal streets enable significant numbers of people to move efficiently and safely through a city. Encouraging more people to walk, cycle, and use public transport as an alternative to using private cars can significantly increase the total network capacity. In doing so, it is also important to design to enable convenient transfer from one travel mode to another. In turn, this will reduce commuting times and lower carbon emissions, improve people's mental and physical health, and enable more leisure time. Additionally, communities will be strengthened through more interpersonal interaction, such as people meeting each other on the street and enhanced safety through more natural surveillance. Economically, local businesses will also benefit from increased footfall and passing trade as pedestrians, cyclists, and public transit users generally spend more money at local retail businesses than people who drive cars. This in turn will enliven the streets further and make them more vibrant and appealing places (see also the section on Economic Value) (Figures 4.32–4.34).

Figure 4.32 Different materials, textures, and colours can help to demarcate zones for different users whilst also creating pedestrian-priority public spaces (Calle Molina Lario, Malaga, Spain).

Source: Photograph taken by the authors.

In some cities, achieving networks of well-designed multimodal streets will require significant changes to the distribution of land uses and activities, particularly in cities that have experienced rigid land-use zoning. In addition, the layout of existing streets and distribution of space to different users will need to be adapted to promote and encourage changing attitudes towards travel choices. This will include reducing the road space dedicated to cars, widening pedestrian areas, providing dedicated cycle lanes, and more transit stops, as well as facilities to make these transit options comfortable such as seating, shelters, cycle parking and storage, and

COMPONENTS AND PRINCIPLES OF URBAN DESIGN 117

Figure 4.33 Successful streets can be multi-modal and accommodate pedestrians and public transport (Goldsmith Street, Nottingham, UK).

Source: Photograph taken by the authors.

transit information equipment (Global Designing Cities Initiative 2016). This should also be combined with greening strategies to promote biodiversity, well-being, and leisure activities. Whilst there are financial costs associated with these changes, the longer-term maintenance benefits of improved quality of life, a healthier and inclusive city, and economic growth will significantly outweigh these.

Figure 4.34 Dedicated cycling infrastructure can encourage safe and convenient travel by bicycle (Cykelslangen, Copenhagen, Denmark).

Source: Photograph taken by the authors.

FURTHER READINGS

Bruntlett, M. & Bruntlett, C. (2018) *Building the Cycling City: The Dutch Blueprint for Urban Vitality*, Island Press: Washington DC.

Higashide, S. (2019) *Better Buses Better Cities: How to Plan, Run, and Win the Fight for Effective Transport*, Island Press: Washington DC.

Newman, P. & Kenworthy, J. (2015) *The End of Automobile Dependence: How Cities are Moving Beyond Car-Based Planning*, Island Press: Washington DC.

Sadik-Khan, J. & Solomonow, S. (2017) *Street Fight: Handbook for an Urban Revolution*, Penguin Books: New York.

Verkade, T. & te Brömmelstroet, M. (2022) *Movement: How to Take Back Our Streets and Transform Our Lives,* Scribe Publications: London.

SOCIAL PRINCIPLES OF URBAN DESIGN

Significantly, the success of public spaces does not depend on size or scale but rather on people's experiences and perceptions of the place. Indeed, activity in the public realm is essential for its vitality and visual attraction, with a critical mass of people helping to ensure that places are enjoyable, safe, and viable. As such, a primary concern of urban design should be a focus on human-centric design, or putting people first, in the creation of successful public places. To ensure that urban areas become multi-use destinations that will attract a diverse range of people for a multitude of reasons to come together for a shared public experience, it is imperative that urban designers understand who the potential users of the public realm are, together with their needs and expectations. The social principles in this section are organised into users; comfortable and enjoyable; healthy and environmentally responsive; safety and natural surveillance; and inclusive and accessible.

USERS

A city's streets and squares, depending upon their cultural, regional, or climatic context, have traditionally been the focus of public life, not just of circulation, but also economic trade, festivals, celebrations, performances, public gatherings, and demonstrations. Indeed, the public realm as a fusion of culture, commerce, and community has provided a dynamic heart of cities for centuries. In the early twenty-first century, a city's public spaces and the activities hosted within them together with the variety of users are more diverse than ever. As such, for cities to thrive socially and economically, it is essential that urban designers aspire to create places that will encourage and promote a truly diverse mix of people to use the public realm. In discussing the relationship between the quality of public space and use, Gehl (1996) suggests that activities in public spaces can be simplified into three categories: *necessary* activities; *optional* activities; and *resultant or social* activities (see Chapter 2). He then argues that many places will have necessary activities regardless of the quality of the place; however, where people have choices, optional and resultant activities, they are much more likely to choose to undertake these in better quality environments. This, therefore, is a good indicator of the types and qualities of places

Figure 4.35 Dancing in public is a popular activity that attracts a diverse range of people and enlivens the street scene in some cultures (Shanghai, China).

Source: Photograph taken by the authors.

that people will make positive decisions to use and that also attract a wide variety of different users (Figures 4.35–4.37).

The types of users and the overall volume of people in any space will, however, also depend on many variables such as the time of day, the weather, the size of the place, how accessible and connected it is, the uses and activities that are present, and the available facilities. It is important to note that not all great places will serve all of society, at least not all of the time. For example, when an event aimed at a particular audience is being held, then others may choose to avoid visiting or using the space. Also, people's

COMPONENTS AND PRINCIPLES OF URBAN DESIGN   121

Figure 4.36 Water features can enhance urban environments and also encourage play and interaction (Taikoo Li Sanlitun, Beijing, China).

Source: Photograph taken by the authors.

Figure 4.37 Spontaneous activities such as street performers enliven spaces, create visual interest, and attract visitors (South Bank, London, UK).

Source: Photograph taken by the authors.

expectations and requirements will differ depending on factors, such as their age, gender, sexual orientation, religion, background, socio-economic group, and education. It is important, however, that all people, groups, and communities have choices and accessibility to public places that meet their needs and desires within all neighbourhoods, towns, and cities. It is also important to realise that people's needs vary at different times of day and night, in different seasons, and in different weather or climatic conditions. In defining people in public spaces, Gehl (2010) suggests that there are five different user categories: *everyday users* who live, work, or study in the area; *passersby* who are passing through or coming and going from other places; *recreational visitors* visiting because of the attraction of the place; *customers* who come from other areas; and *event visitors* who are attracted by a particular event. The most successful places are likely to attract as many of these user groups as possible on a regular basis.

FURTHER READINGS

Gehl, J. (1996) *Life Between Buildings*, Island Press: Washington DC.
Gehl, J. (2010) *Cities for People*, Island Press: Washington DC.
Gill, T. (2021) *Urban Playground: How Child-Friendly Planning and Design Can Save Cities*, RIBA Publishing: London.
Matan, A. & Newman, P. (2017) *People Cities: The Life and Legacy of Jan Gehl*, Island Press: Washington DC.

COMFORTABLE AND ENJOYABLE

People need to want to use the public realm for it to be successful, and therefore, the design of all public spaces needs to take human comfort and enjoyment into account. This includes facilitating people's daily needs such as moving between home and other destinations such as work, school, and shops but also creating opportunities for people's leisure time including places to sit, relax, meet others, to play, etc. (Sim 2019). Therefore, in addition to providing space for various activities to happen, there needs to be appropriate and sufficient infrastructure and facilities in place to encourage people to maximise the opportunities that public space can offer.

These include appropriately designed street furniture, such as seating and tables, in convenient and desirable locations, together with shelter from rain, snow, wind, or sun depending on the specific climatic context.

The design and locations of these need to be considered carefully as part of the whole design of the public realm to ensure that they are appropriate and support the patterns of street activity. In addition, the provision of contemporary needs such as cell phone and laptop charging stations can facilitate people's day-to-day lives. Careful consideration should also be given to the choice of fixed or portable furniture depending on the anticipated use of the space. Indeed, in public places where more adaptability is required to enable a more varied range of events, then flexible or portable furniture, including suitable storage facilities, will not only enable different events and activities to be accommodated but also enable people to personalise their experience of the space. When achieved successfully, public spaces will attract people if they are comfortable, enjoyable, and able to meet the demands of different users at different times (Figures 4.38–4.41).

It is also imperative that urban designers follow a human-centric approach to placemaking. This involves *"...creating a human scale environment means making sure that the objects that we interact with every day are of a size and shape that is reasonable for an average person to use"* (Burke 2016). Indeed, if people are to feel comfortable with the public realm, then the design of the streets, public squares, and the surrounding architecture needs to respond to the human scale. Urban design should therefore require human-scale building edges to all public spaces and also arrange street lighting, wayfinding, and signage in relation to the human eye level. Similarly, the design of street furniture and other public facilities should accommodate universal accessibility to ensure its appropriateness for all people and communities.

FURTHER READINGS

Burke, S. (2016) *Placemaking and the Human Scale City,* Project for Public Spaces. www.pps.org/article/placemaking-and-the-human-scale-city

Gehl, J. (2010) *Cities for People*, Island Press: Washington DC.

Figure 4.38 Steps create important places for people to rest, to eat and drink, to meet friends, and to people watch within cities (Église de la Madeliene, Paris, France).

Source: Photograph taken by the authors.

Project for Public Spaces (2023) *What Makes a Successful Place?*. www.pps.org/article/grplacefeat.

Sim, D. (2019) *Soft City: Building Density for Everyday Life*, Island Press: Washington.

HEALTHY AND ENVIRONMENTALLY RESPONSIVE

> *Urban areas need to be planned more effectively to enhance their benefits and reduce the threats to healthy development.* (RTPI 2014: 9)

In the face of climate change, urban designers are increasingly concerned with approaches that promote environmental sustainability in the creation of eco-friendly and resilient urban environments. Indeed, improving the design of our cities and their public realms can, not only have a positive impact locally but also, help to address important wider agendas such as greenhouse gas emissions, and global warming. Whilst sustainable movement

Figure 4.39 Formal and informal seats, such as low walls, can make excellent places for people to rest, meet friends, and people watch (Suzhou, China).

Source: Photograph taken by the authors.

Figure 4.40 Opportunities to sit alongside water provide an attractive location within cities (Rathaumarkt, Hamburg, Germany).

Source: Photograph taken by the authors.

Figure 4.41 Steps and seating linked to buildings can create public amphitheatres for congregating, watching events, and observing people (Sea World Culture and Arts Center, Shenzhen, China).

Source: Photograph taken by the authors.

and transit approaches (see also the sections on Walkability, and Multi-Model Transit) can have a significant impact on this, it is also imperative that streets and squares are designed to support healthy environments and lifestyle choices (Project for Public Spaces 2022). Indeed, integrating green infrastructure strategies into the design of public realm can improve air and water quality, encourage physical activity, reduce stress, and improve mental health (Figures 4.42–4.44).

The urban micro-climate is therefore an important consideration in urban design projects. Importantly, promoting contextually sensitive green infrastructure, such as native species of trees and

Figure 4.42 The combination of water and greenery creates an attractive urban environment, especially in warmer climates (KLCC, Kuala Lumpur, Malaysia).

Source: Photograph taken by the authors.

plants, can help to significantly improve air quality, provide shade, improve water management systems, promote biodiversity, and support local ecosystems. In doing so, design solutions should be informed by and complement existing natural features and habitats, climate, topography, and water bodies. Indeed, appropriately located street trees and landscaping not only improve the visual amenity and therefore attractiveness of a place but also contribute positively to improving the local climate. This includes helping to reduce the urban heat island effect, thereby making the urban environment more comfortable and reducing the cooling requirements of adjacent buildings. In addition, urban trees can help to improve air quality and reduce noise pollution. Similarly, well-designed water features not only provide an attractive amenity for users of the public realm to engage with but can also help to improve drainage, create pleasant sounds, help to make people feel cooler, and contribute to a community's well-being.

128 URBAN DESIGN

Figure 4.43 Waterside places with shade from trees can provide attractive locations for cafes and restaurants (Suzhou, China).

Source: Photograph taken by the authors.

FURTHER READINGS

Barton, H., Grant, M. & Guise, R. (2021) *Shaping Neighbourhoods for Local Health and Global Sustainability*, Routledge: London.

Bentley, I., De, S., McGlynn, S. & Rampuria, P. (2024) *Eco Responsive Environments: A Framework for Settlement Design*, Routledge: London.

Gehl Institute (2017) *Inclusive Healthy Places: A Guide to Inclusion and Health in Public Space – Learning Globally to Transform Locally*, Gehl Institute with the Robert Wood Johnson Foundation. https://ihp.gehlpeople.com/wp-content/uploads/2022/08/Inclusive-Healthy-Places_Gehl-Institute.pdf (accessed 29 November 2024).

Montgomery, C. (2015) *Happy City: Transforming Our Lives Through Urban Design*, Penguin: London.

Figure 4.44 Canal networks and rivers with their distinctive characteristics create unique and attractive contexts in many towns and cities (Groenburgwal, Amsterdam, Netherlands).

Source: Photograph taken by the authors.

Project for Public Spaces (2022) *Placemaking: What if We Built Our Cities Around Places: A Placemaking Primer*, PPS: New York.

Roe, J. & McCay, L. (2021) *Restorative Cities: Urban Design for Mental Health and Wellbeing*, Bloomsbury: London.

SAFETY AND NATURAL SURVEILLANCE

Natural surveillance is critical to people's perception of, and actual experience of being safer when using the public realm. Indeed, well-observed and overlooked places significantly limit the opportunities for crime by increasing the chances of offences being observed. Often referred to a 'eyes on the street', people feel safer and have perception of safety when there are other people around who can see what is happening and have the potential to intervene if any incident was to occur. Generally, the more people within a public space the safer it feels unless it actually becomes too overcrowded. In addition to other users within public places contributing to the feeling of safety, people with views of the space from the ground and lower floors of neighbouring buildings can also contribute to the natural surveillance of the space. Ironically, even people in passing cars, if driving at slow speeds, can offer some natural surveillance in the public realm.

Successful urban design can improve the natural surveillance of public places in a number of ways. These include the creation of places that will have many users undertaking a wide range of different activities at all times of the day and night. Achieving this will require a good mix of land uses that will attract diverse users for different reasons and at different times (see also the section on Mixed-Use and Multi-Dimensional). In addition, there are physical aspects to consider when designing public spaces, such as: having good visibility and sight lines from adjacent buildings; ensuring active edges with visual and physical activity; maximising visibility by avoiding blind spots; carefully locating potential visual obstructions such as street furniture or landscaping; and providing effective artificial lighting.

People feel more comfortable using safe public spaces and congregate where there is a sense of protection against unpleasant or unsafe experiences. The public realm should therefore be designed to be safe and comfortable for all users; however, the focus should be on the needs of pedestrians and cyclists. In particular, the most vulnerable users such as children, seniors, and people with disabilities should be prioritised. Design approaches should therefore reduce people's exposure to potential conflicts, provide natural surveillance, and ensure spaces are safely lit, and free of hazards. As such, public spaces should, where appropriate, be designed for slower traffic speeds and include wide pavements, and designated cycle paths, with high-quality lighting and street furniture, to provide a safe user experience and support crime prevention.

FURTHER READINGS

Jacobs, J. (1961) *The Death and Life of Great American Cities*, Random House: New York.

Oc, T. & Tiesdell, S. (1997) *Safer City Centres: Reviving the Public Realm*, Chapman: London.

Sim, D. (2019) *Soft City: Building Density for Everyday Life*, Island Press: Washington DC.

INCLUSIVE AND ACCESSIBLE

The public realm has a pivotal role to play in enriching a city's social fabric by providing places where communities can congregate, socialise, and participate in activities. Careful and considered design and management of the public realm is therefore paramount, not just for the physical appearance and economic success of a place, but also for fostering a city's social and cultural dynamics. Indeed, if well designed, the streets and squares can enhance and enrich the societal fabric of a city. It is imperative, therefore, that the design of the public realm creates an inclusive environment relevant to its socio-cultural context. In doing so, all spaces within a city should be accessible to all, regardless of age, gender, mobility, or background. Indeed, Pineo (2022: 983) argues that: *"An inclusive built environment enables all members of society to conveniently participate in daily activities without feeling that they are disadvantaged by their personal characteristics or needs."*

Whilst well-designed and successful public spaces can enhance people's quality of life by providing places for relaxation, recreation, and enjoyment, they also play a key role in enhancing well-being in urban areas. Empowerment and social inclusion should therefore also be central to the design decision-making process to create inclusive places for all of society's diverse user groups including the most vulnerable. Indeed, well-designed public spaces are important for connecting people with their communities, providing opportunities to meet others, and enabling a feeling of being socially connected. Achieving this requires all aspects of the place to be accessible and welcoming from the general atmosphere to the materials, facilities, and landscape. Spaces should therefore be designed with an awareness of the needs of diverse users to ensure everyone can participate equally within the public realm. This involves the integration of universal design principles to create spaces that promote social interaction amongst all members of society to provide places that are attractive, functional, and celebrate the spirit and culture of local communities. Importantly, achieving inclusive and accessible public places will be a significant catalyst for social cohesion and community identity (Figures 4.45 and 4.46) (Project for Public Spaces 2022).

Figure 4.45 Temporary activities help to transform public spaces and attract different users (Festival City, Dubai, United Arab Emirates).

Source: Photograph taken by the authors.

FURTHER READINGS

Burton, E. & Mitchell, L. (2016) *Inclusive Urban Design: Streets for Life*, Routledge: London.

Flanders, D. & Miller, E. (2020) *Creating Great Places: Evidence-based Urban Design for Health and Wellbeing*, Routledge: New York.

Gill, T. (2021) *Urban Playground: How Child-Friendly Planning and Design Can Save Cities*, RIBA Publishing: London.

Lofland, L. (1998) *The Public Realm: Exploring the City's Quintessential Social Territory*, Routledge: London.

Shatoe, H. (2008) *Convivial Urban Spaces: Creating Effective Public Places*, Earthscan: London.

Figure 4.46 Pop-up activities such as bookstores can transform underused spaces within cities (South Bank, London, UK).

Source: Photograph taken by the authors.

Thwaites, K., Mathers, A. & Simkins, I. (2013) *Socially Restorative Urbanism: the Theory, Process, and Practice of Experiemics*, Routledge: London.

Tonkiss, F. (2013) *Cities by Design: The Social Life of Urban Form*, Polity Press: Cambridge.

OPERATIONAL PRINCIPLES OF URBAN DESIGN

A city's public spaces are the intersections of community, commerce, and culture. The public realm, like cities themselves, is continually evolving; however, its role as the centre of public life is essential for success, sustainability, and resilience of urban life. Achieving this requires that all of a city's public spaces are well-designed, maintained, and managed. Inevitably, some of these responsibilities lie with various authorities who impact upon the on-going organisation and operation of these public places. Where this is done successfully, and communities are actively engaged in these processes, there can be considerable benefits for the public realm. This section will therefore focus on the management and maintenance of public space, the importance of public engagement and participation in these processes, and how the public realm can add economic value to the city.

MANAGEMENT AND MAINTENANCE

A city's public realm needs to be well managed and maintained to ensure that it continues to be successful and serve the needs of its users. Importantly, this needs to be a continuous process that is responsive to the changing and evolving requirements of the communities that engage with the public spaces. As such, cities will require detailed maintenance plans if the public realm is going to continue to attract the public to use the space. Whilst good design can help to minimise the on-going maintenance, from time-to-time, infrastructure, utilities, surfaces, and facilities will need cleaning, repairing, upgrading, and replacing. Litter and refuse will also need collecting and recycling and landscapes will need regular and seasonal maintenance. In addition, it is important during the design process to consider how the public realm will facilitate emergency services during their daily use and when events are occurring in public spaces.

Strong management strategies also help to ensure the longevity and sustainability of successful streets and squares. This includes not only public transportation but also curb-side management strategies for deliveries, access, drop-off, and parking together with relevant pricing strategies. Increasingly, city authorities are also adopting place management strategies to ensure regular programmes of activities to animate and enliven the public realm. This can include the planning, organisation, and implementation of events such as markets, fairs, festivals, performances, and other attractions that will encourage both local people and visitors to use and spend time within a city's public spaces (Figures 4.47 and 4.48).

FURTHER READINGS

Carmona, M., Bento, J. & Gabrieli, T. (2023) *Urban Design Governance: Soft Powers and the European Experience*, UCL Press: London.

COMPONENTS AND PRINCIPLES OF URBAN DESIGN 135

Figure 4.47 Urban spaces that are badly designed, with no purpose, and poor maintenance discourage users and therefore feel unsafe and impact negatively upon the local environment.

Source: Photograph taken by the authors.

Figure 4.48 Strong management of public places can include seasonal events with temporary attractions that entice many visitors (Paris Plage, Paris, France).

Source: Photograph taken by the authors.

Hayward, R., McGlynn, S. & Reeve, A. (2010) *Better Towns and Cities: A Manual of Town Centre Management*, Architectural Press: Oxford.

Punter, J. (ed.) (2010) *Urban Design and the British Urban Renaissance*, Routledge: London.

PUBLIC ENGAGEMENT AND PARTICIPATION

It is essential that people-oriented approaches are adopted for all stages of the urban design process and the ongoing delivery and management of public spaces. Importantly, the cooperation and participation of users is crucial not only in building and strengthening local communities but also in the achievement of an inclusive and successful public realm. Indeed, the active engagement of all local stakeholders, people, and organisations, can play an essential role in effective decision-making that represents the interests of all users of a city's public spaces. When undertaken in an equitable and impactful manner, such participation can help to strengthen communities, ensure relevant places, and foster a sense of belonging and ownership for local people. Such an approach also helps to ensure that invaluable local knowledge and experiences can inform

Figure 4.49 The most successful public spaces tend to be inclusive of, cater for, and attractive to all members of society regardless of age, gender, race, etc. (Old Market Square, Nottingham, UK).

Source: Photograph taken by the authors.

Figure 4.50 Rooftops, even those that are privately owned, can become seasonal social spaces and offer alternative places and views for visitors (Galeries Lafayette, Boulevard Haussman, Paris, France).

Source: Photograph taken by the authors.

the design process and on-going use of the public spaces and therefore help to ensure the success of projects. The engagement of local communities and other stakeholders should be undertaken using a range of participatory processes to ensure as wide a participation as possible from all stakeholders (Figures 4.49 and 4.50).

FURTHER READINGS

Malone, L. (2019) *Desire Lines: A Guide to Community Participation in Designing Places,* RIBA Publishing: London.

Natarajan, L. & Short, M. (2023) *Engaged Urban Pedagogy: Participatory Practices in Planning and Place-Making,* UCL Press: London.

ECONOMIC VALUE

A safe, vibrant, efficient public realm consisting of a network of streets and squares is essential to the economic and social health of a neighbourhood or a city (Carmona 2019). Indeed, a city's public

Figure 4.51 Places that continue to attract people at all times of day, on all days of the week, and at all times of the year tend to be well-managed to ensure that they are clean and safe with a range of varied activities to cater for all interests (Rambla de la Llibertat, Girona, Spain).

Source: Photograph taken by the authors.

places can be an important economic asset as well as a functional element. A well-designed public realm can also help to stimulate local economies by attracting locals and tourists thereby encouraging the incubation of new businesses (Project for Public Spaces 2022). These economic activities help to support livelihoods, create value, and foster economic development and growth. In simple terms, well-designed streets create urban environments that encourage people to stay and spend time, generating higher revenues for traders and local businesses, more job opportunities, higher values for property owners, and more taxes for local governments (see also the section on Multi-modal Transit) (Figures 4.51).

A well-designed public realm network can also provide many additional cost benefits for a city, its businesses, and its communities. An efficiently planned city can facilitate walking, cycling, and public transport use and save people considerable time in going about their daily lives. In addition, there are significant financial savings in terms of reduced fuel costs, less need for private vehicle ownership, fewer accidents and injuries, and less work hours lost due to congestion. More importantly, adopting more sustainable ways of moving around cities will lead to improved physical and mental health and wellbeing which will reduce medical expenses and the demand on healthcare and social services. Other positive economic impacts include a reduction in environmental costs such as pollution and well-designed streets and squares can also help to deal with climatic costs such as flooding by utilising sustainable stormwater and drainage solutions.

FURTHER READINGS

Alvarez, L. (2023) *A Beginner's Guide to Urban Design and Development: The ABC of Quality, Sustainable Design*, Routledge: New York.

Carmona, M. (2019) Place Value: Place Quality and Its Impact on Health, Social, *Economic, and Environmental Outcomes, Journal of Urban Design*, 24(1): 1–48. https://doi.org/10.1080/13574809.2018.1472523

Larco, N. & Knudson, K. (2024) *The Sustainable Urban Design Handbook*, Routledge: New York.

Savours, B. (2024) The Planning Premium: The Value of Well-made Places, June, Public First Limited: London. www.publicfirst.co.uk/wp-content/uploads/2024/06/The-Economic-Value-of-Good-Town-Planning-Final-1-1.pdf

SUMMARY

This chapter has examined three fundamental considerations for urban design practice in most contexts around the world. Firstly, the characteristics of successful public spaces were reviewed to enable an understanding of what qualities can help to contribute to the creation of a public realm that is appropriate for its context and works well for all users of the space. It is important to emphasise, however, that not all, indeed relatively few, great public places will display all these qualities. Nevertheless, those places that are more successful are likely to demonstrate many of the characteristics highlighted and some of these can contribute for the lack of others. It is, however, important that in the design or refurbishment of public spaces, that the goal is to achieve as many as possible to help ensure the quality and resilience of any specific public space. These qualities of successful places also need to be considered and adapted in relation to specific contexts such as the social, cultural, and climatic characteristics of the place to ensure the contextual suitability of the project.

The second main section focused on the key physical components of urban areas and their role in organising and creating the public realm. Indeed, the three fundamental components, the street, the public square, and the urban block, as explained are the fundamental physical ingredients of a successful urban environment. All of them have historic precedents that date back through the evolution of settlements and have, in different forms, provided successful global examples of successful urban design practice around the world. The section also briefly discussed the importance of the international and cultural context in terms of how these elements might be appropriate and applied in various contexts.

The final section considers fundamental principles that should be considered during the urban design process. Again, these can be applied across different contextual situations, however, the way that they might be applied will vary for every place in order to achieve appropriate, resilient, and sustainable solutions. Nevertheless, all of these principles, which are multi-dimensional, interrelated, and often cannot be considered in isolation, need to be considered to ensure the creation of high-quality public spaces and that the public realm becomes an enduring and vibrant part of urban townscapes.

THEORIES ON SUSTAINABLE URBANISM

This chapter covers a brief introduction to the theories of urbanism, including some macro perspectives on exploring human perception and behaviour in urban spaces. The main objective is to explore the factors needed for a more sustainable form of living in our cities. For urban designers, it is a basic requirement to understand how built spaces affect humans and their everyday spatial practices to assess their physical configuration to reduce adverse effects with design interventions. The scientific field of urbanism itself can be identified as interdisciplinary and located between social sciences (e.g. sociology, human geography, and environmental psychology) and urban design as an applied science focusing on the built environment and its effects on everyday life.

We first start with an overview on the terminology of urbanism as discussed by Louis Wirth and the Chicago School of Sociology at the beginning of the twentieth century. This is followed by a small philosophical discourse revisiting the theory of space production introduced by Henri Lefebvre and his methodology of rhythm analysis to explore the impact of urban spaces and their representations on everyday urban life. We then continue with an exploration of key qualities needed in the built environment to support the reduction of wasting natural resources and to support a more sustainable development. Sustainability is explored in this chapter following the general definition of the United Nations Brundtland Commission from 1987: *"Meeting the needs of the present without compromising the ability of future generations to meet their own needs."* Any urban design project is ideally in line with national and regional strategies to reduce the waste of resources, which is rooted in the way we are living in cities, particularly referring to our housing and mobility choices.

As an applied science, urban design is dependent on social sciences and their various research outputs within urban studies since the core objective of any urban design intervention is to improve the quality of urban life. It is therefore important to differentiate the various fields of social sciences before moving on to explore the distinctive theories and their methodological approaches. While humanities and natural sciences have evolved as main fields of human knowledge for centuries, social sciences have often been explored as an adjunct part of humanities: as a pragmatic discourse on observing societies and their behaviour within a historic context and often initiated to inform governance (Case & Vanderweele 2024). With the emerging impact of natural sciences during the Age of Enlightenment in surveying and understanding human behaviour, social sciences have slowly established their own scientific traditions between both fields. Nevertheless, it is important to note that the works of moral philosophers and historians must be regarded as the starting point of knowledge acquisitions how societies form and behave in specific environments (Coates 2023).

Whilst humanities focus on human beings and their creative outputs (including languages) as unpredictable phenomena, natural sciences have emerged to draw a more quantifiable and deterministic picture of human beings as being another dependent part of the natural world. As a synthesis of both traditions of knowledge inquiry, social sciences can be criticised to neither reach out to the depths of a philosophical and historical discourse nor to the detailed biological and functional principles of the human organism. In many respects, social scientists have thus often been disregarded as merely being observers and reporters of human behaviour with less valuable theories on human motivations and less valuable insights into scientific systems (Holmes 1997). The importance of social sciences can be mainly observed with the rise of modern capitalist societies and increasingly complex social structures and their conflicts requiring more critical observations and derived theories on human behaviour and perception. Subsequently, social sciences have become an important institutionalised scientific field to understand development trends and their factors in various contexts to inform new approaches to modify behavioural patterns, via policies, marketing, or strategic investments.

The various fields of social sciences can be organized based on their main methodological approaches: First, geography is an important discipline, which is often referred to as being on the junction between sciences and social sciences. In contrast to other social sciences, the main objective of geography is to quantitatively survey our planet which includes the various forms of how human behaviour has been impacting natural elements and resources and their distribution. The main research questions are thus rooted in exploring the effects of human behaviour rather than its origin, and the methods include various approaches in mapping surveys. This macro-scale is rather limited in introducing new theories on human behaviour. Thus, it usually follows deductive approaches to collect and assess data on how cities form, use resources, and impact the natural environment. Most geographic sciences, including urban geography, are therefore centred on monitoring human behaviour and to understand the functional aspects of economic networks and the associated urban development. Human geography on the other hand is a field which combines quantitative and qualitative investigations, since its scale is significantly smaller, and its main objective is to study socio-spatial relationships of human beings and their environments (Campbell 2018).

The scientific lines between human geography and sociology intersect due to the strong relationship between both fields in studying human–environment interactions. While human geography is generally more concerned to understanding how spatial and environmental factors influence human behaviour, sociology is more concerned on social structures and institutions. The rather blurry transition from human geography to urban sociology has led to the rather broad recognition of various scientific traditions being found in the interdisciplinary and open field of urban studies. This chapter will thus commence with an excursion to the Chicago School, one of the pioneers of modern urban sociology and anthropology, which has also affected human geography and its main objective to survey spatial practices (Ohm 1988). The dynamic relationship between geography and sociology can be seen as important starting point to reflect on the emergence and role of social sciences, since it clearly illustrates the two scientific departures of either quantifying the human impact within our system of space and time or to qualitatively assess social structures to start a discourse on human behaviour and its motivations. Since both quantitative and qualitative methods are needed to understand any city and its development, urban and human geography as well as urban sociology, and urban anthropology can be summarised as urban studies, which also include various other fields, such as urban economics and political sciences focusing on the behaviour of markets and governance.

Following this discourse on the origin of urban studies at the beginning of the twentieth century, Henri Lefebvre, and his theory on the production of social space need to be introduced to reconnect to the nineteenth century's tradition of dialectical philosophical thinking and thus a more holistic framework to understand spatial practices. The increasing interests in human behaviour and its factors have also affected the further development of psychology to widen its field, as in the case of environmental psychology. This leads to a broader discussion on the imageability and language of urban spaces, rooted in the understanding of our shared ways of how we perceive spatial configurations to explore key factors of our spatial orientation and practice. The historic formation of many branches of social sciences can be regarded as a natural process of specialisations throughout history to understand societies from various viewpoints. All these branches meet in the case of cities and can thus be relevant to both urban planning and design decisions. It is thus important to note that the relevant literature for urban designers is expanding to all outputs of social sciences investigating human behaviour in urban contexts and only selected key theories and approaches can be briefly revisited in this chapter.

LOUIS WIRTH AND HIS THEORY ON URBANISM

Before we can explore the theoretical frameworks of space production and sustainable urbanism, the general roots of studying human perception and behaviour in urban spaces need to be introduced. One of the main initial motivations was the categorisation of behavioural patterns linked to the various social groups within a complex and changing urban population. This area has been the topic of first sociologists and human geographers, especially in the context of the United States due to complex urban growth dynamics and the big variety of migrant groups. One of the first influential articles defining the widespread use of urbanism as a scientific term in the English language was written by Louis Wirth for the American Journal of Sociology in 1938: *Urbanism as a Way of Life*. He postulated his theory defining urbanism as the study of how urban populations live in cities by identifying key factors. In doing so, Wirth revisited the discourse on the city and its meanings for society by Max Weber (1930) and Robert E. Park and Ernest W. Burgess (1925) to establish a more coherent theoretical framework and a new basis for empirical research on urbanism. At first, he noted that there are three critical parameters defining urbanism from a sociological standpoint: (a) the population size; (b) the density of the urban settlement; and (c) the heterogeneity of the social group (Wirth 1938: 10).

In his view, there was sufficient empirical evidence that an increased number of inhabitants would lead to an enhanced differentiation between individuals, which is expressed in a bigger cultural diversity in cities in comparison to rural communities. This tendency would also explain the trend of spatial segregation and social enclaves in cities to protect cultural peculiarities, while there is a general trend of social disorganization due to fragmented interpersonal relationships (Wirth 1938: 11 and 13). Louis Wirth recommended to study urbanism by integrating three basic dimensions: (a) the physical structure of the urban environment and its population density; (b) the system of social organization, including governance and typical patterns of social relationships; and (c) a set of ideas, expressed in collective behaviour (Wirth 1938: 19). The assessment of the spatial configuration of housing typologies and the resulting urban densities is thus one of the main dimensions defining the way we live in our cities, while the social structure and socio-cultural dimension need to be seen as equally important. Subsequently, different social groups would interact in some respects differently in the same spatial configuration: For instance, a low-income social group with less leisure time and financial means would lead to less populated public spaces and less diversity in local commerce.

Overall, Louis Wirth was convinced that living in a city changes the way people behave and interact. Due to the complexity of social backgrounds and varying densities, we can observe a big diversity of social life in any bigger city. One of the main challenges is to identify key forms of urbanism since they are always facing change due to continuous developments. He also observed an increasing lack of personalness and communication due to more social interactions between strangers for only short periods of time. On that basis, Louis Wirth postulated his hypothesis that urban dwellers are more inclined to status symbols, expressed in external appearances, and the pursuit of personal happiness in form of consumption (Wirth 1938). Another key aspect in this context is the increasing level of mobility shaping a new reality of less stable social networks expressed in local communities. One consequence is the need for more formal social control in urban societies, whilst in rural communities, family ties and customs are often observed as more effective. The main reasons of weaker community bonds in cities need to be seen in the fact that most city dwellers have not lived together for generations. One threat, according to Louis Wirth, is the resulting rationalised behaviour of these dwellers causing less willingness to participate and to claim responsibility (Wirth 1938).

His statement of seeing urbanism as a social organisation endangering the culture of smaller social groups due to more secondary than primary social contacts has led to the rather common criticism of Louis Wirth's hypotheses to be too grounded in his own limited observations, especially in Chicago itself. Furthermore, human relationships in cities are complex and it can be expected that a much bigger quantity and frequency of social interactions can be expected in cities than in rural areas, making the comparison to rural areas dependent on actual numbers and more accurate observations. Despite the understandable concerns on the rather simplified idea of perceiving urbanism, as the way we are living in cities, defined by population size, density, and the heterogeneity of urban populations, Louis Wirth drew a new attention to the understanding minority groups and the impact of cities on their changing behaviour. The resulting urban lifestyles can be studied and compared, which has contributed to the evolving scientific fields of urban sociology and anthropology, which have had a significant role

Figure 5.1 A gentrified neighbourhood with refurbished buildings (Notting Hill, London, UK).

Source: Photograph taken by the authors.

in supporting urban design decision-making to be more responsive to the local social context and to be more aware of the future impact of new developments, such as gentrification trends (Figure 5.1).

Beyond this essential view on his theory, Louis Wirth's contributions need to be understood in the context of the *Chicago School of Sociology*, which is regarded as one of the first pioneering centres of urban sociology and anthropology. The core theories were formed by various other researchers, such as Robert Park and Ernest Burgess. One specific aspect of the Chicago School was the emphasis on micro-scale social interactions in cities, including George H. Mead's theory on pragmatism and symbolic interaction emphasising the importance of studying people's actual activities and interactions, as in his book *Mind, Self, and Society* (1934). One of the biggest motivations of the Chicago School was to inquire answers on the key factors of social problems, as the result of the rapid urbanisation during the second Industrial Revolution in the United States. Chicago itself witnessed a dramatic growth from a small town of around 10,000 inhabitants in 1860 to a big urban agglomeration with around two million inhabitants 50 years later (Burgess 2008). Both the rapid urban growth and the big diversity of migrant groups formed an important testing ground for many of the School's first theories.

The continuous relevance of these theories has often been overlooked due to the specific historic period and location of the School itself and the missing reflection on its new methodological approach how to study urbanism. According to Andrew Abbott (1997), the most important contribution of Louis Wirth and his colleagues has been the observation that the urbanism or general behaviour of certain social groups depends on their specific local context of time and place and cannot be reduced to only a few factors. Their original approach of observing both the behaviour of social groups and their distinctive social and spatial settings was challenged by modern surveys and the focus on statistics to identify causalities between certain variables and social behaviour in bigger scales (Abbott 1997). While identifying variables can be an important strategy to assess general behavioural trends, the complexity of behavioural patterns needs to be studied in the specific context of time and place to inform decision-making. As Abbott (2020) points out, the Chicago School has often been considered to have lost its connection to local city planning once it moved to a focus on delivering big-picture theories on urbanism and an associated generalisation of variables impacting human behaviour.

Louis Wirth had an important role as being part of the National Planning Board, which was an interdisciplinary group submitting reports in several stages from 1937 to 1939 (Wirth 1938). He and his research assistant Edward Shils contributed detailed research on housing dynamics in relation to social class including the physical aspects of the various housing typologies and their impact on urbanism via various case studies. An important characteristic of this report was the focus on critical observations rather than policy solutions. The report was centred on observations of people in all main urban settings in the United States and investigated the relation between variables, such as city size, population growth, and housing supply. The main objective of this report was to identify aggregate properties to predict development patterns, and it was thus already affected by more quantifiable aspects and less particular local urban settings. In many respects, Wirth's journal article "Urbanism as a Way of Life" can be seen as his attempt during the late 1930s to move away from particular social groups to a generalisation of urban phenomena. This move from a smaller scale to bigger theories on urbanism can be seen as critical to understanding the limited role of the School in decision-making at a local scale (Wirth 1938).

In summary, urban sociology can be regarded as a key scientific field to study for any urban designer, since the one shared objective of any design intervention is to improve the ways, we live in our cities. While new methodologies were added since the 1930s, some core approaches have remained. Qualitative observations of public spaces and semi-structured interviews with social groups were applied by first urban sociologists and anthropologists and are still applied today to study how spatial settings affect human behaviour and perception. In addition, the quantitative surveys of both the built environment and demographic data have been key to supporting planning decisions, which have affected urban design concerns, such as the required urban densities and preferred housing typologies as well as public spaces and mobility modes (Figure 5.2).

Figure 5.2 The new transit-oriented district of Riedberg with shared green spaces in the urban periphery of Frankfurt am Main (Germany).

Source: Photograph taken by the authors.

FURTHER READINGS

Gottdiener, M., Budd, L. & Lehtovuori, P. (2015). *Key Concepts in Urban Studies*, Sage: New York.

Mumford, L. (1961) *The City in History. Its Origins, Its Transformations, and Its Prospects*, Harcourt: New York.

Philips, B. (2010) *City Lights: Urban-suburban Life in the Global Society*, Oxford University Press: Oxford.

White, W. (1980) *The Social Life of Small Urban Spaces*, Project for Public Spaces: New York.

HENRI LEFEBVRE AND HIS THEORY ON SPACE PRODUCTION

After the various explorations on urbanism by the Chicago School of Sociology in the context of the 2nd Industrial Revolution, there has been a significant rise of research on urbanism and its theoretical grounds in the post-war period. This time is often referred to as the first urban crisis, when car-centred urban planning took charge resulting in urban sprawl and land use following the new rationale of spatially dividing urban functions and establishing modern infrastructure. The French philosopher Henri Lefebvre had a major impact on our today's understanding of this significant period in its historical context by discussing the dialectics of socio-spatial interactions. Despite his academic background being rooted in philosophy, he himself was a big promoter of urban studies to become a more mature and independent subject to foster the integration of all perspectives and to overcome the divided landscape of social sciences and its various disciplines. His main approach in investigating urbanism was mainly rooted in the humanist tradition of phenomenology and historical discourses. He was purely interested in a qualitative debate following the traditions of dialectical thinking, as initiated by Georg W. F. Hegel, and continued by Karl Marx and Friedrich Nietzsche towards the end of the nineteenth century (Schmid 2022).

This small excursion to a philosopher's perspective on urbanism can be seen as a critical starting point to understanding the dimensions defining the way we live in cities. As one of the first pioneers, Henri Lefebvre discovered everyday urbanism as a phenomenon to be investigated. One main reason can be found in the already mentioned experienced urban crisis in the middle of the twentieth century resulting in new forms of everyday living. Another factor has been his extensive study of Karl Marx and his shared perspective of overcoming the Hegelian idealism and its dependence on an increasingly abstract and detached philosophical tradition. Henri Lefebvre was convinced that the human experience is dependent on both the ever-moving physical world as well as the ideas and thus perspectives drawn from it. In his view, a discourse on urbanism needs to commence with the two opposing poles: the countryside and the city. For centuries, the city was a marketplace formed by the surrounding countryside and the territorial expansion required cities to host the military and its technological advancements, as found in the European antiquity and other civilisations. The main turning point in history was identified by Henri Lefebvre in the case of the European Renaissance and the beginning of complex trading networks challenging feudalistic principles and overcoming them.

In his view, the industrialisation has been a logical consequence of these advancing networks, and based on assuming an on-going evolutionary process, he postulated his idea of an urban revolution (Lefebvre 2002) as the next and final stage of development. Henri Lefebvre's work clearly reflects his understanding of cities being no detached units but instead being the products of two main levels of continuous movement and change: The individual personal and private level rooted in the everyday form of living and the global level of major dynamics in markets and decision-making on bigger scales. In many ways, he was able to predict globalisation trends following the rationale of expanding consumerism and the subsequent conflicts of social spaces: On the one hand, there are new spaces of leisure with increasingly less experiences of manual labour, and on the other hand, the spaces of labour have not disappeared but have moved and concentrated in often segregated places. Housing and urbanism are often mentioned in his influential essay *Le Droit à la Ville* ("Right to the City") from 1968 (published in English in 1996) to underline and explain these conflicts via examples, such as the urban sprawl of Paris in form of low-rise dwellings and mass housing, known as *Banlieus*.

His critical reflections on everyday urbanism are always rooted in his dialectical discourse on how we produce space. He enriched this debate via both his historical excursions and the associated dialectics between urban and rural spaces in different eras and via his dialectics of scale between private and global dimensions. Time and scale have however not been the only reflection on space as a social space. In his book *The Production of Space*, Lefebvre (1991, originally published in French in 1978) introduced our social space in cities as the product of three core dimensions of everyday practices rooted in perceptions. The first dominant socio-spatial interaction has been defined by the instinctual need for survival and thus by the moment-to-moment perception and adaptation regarding environmental conditions. Lefebvre named this ad hoc dimension perceived space, and in its dominant form, it would lead to absolute spaces, which we can still observe in collectively built vernacular settlements, or historic remains defined by intrinsic climatic and cultural conditions (Figure 5.3).

The ability of the human mind to abstract spaces and to plan spatial forms via the collection of empirical evidence led to the first major transformation of urbanism in which few decision makers were able to define new socio-spatial conditions. These conceived spaces led to the implementation of often geometrical and abstract forms implementing a top-down and efficient order of spatial as well as social organization. Based on the historically documented role of elitist social structures in any civilization, conceived and abstract spaces often led to temporary social order and subsequent social conflicts fuelled by the rapid increase in social inequalities expressed in spaces of labour and spaces of leisure (Lefebvre 1991). One core factor in avoiding social unrest threatening the planned socio-spatial condition has always been the complex and subjective place attachment shared between individuals.

Henri Lefebvre discussed this third and final dimension as lived space or space of representations, which has also been investigated by theorists, such as Edward Soja (1996). The everyday individual perception of aspects like safety and comfort meets the need for a historic, cultural, and symbolic representation of belonging. Any attachment and thus long-term investment of an individual in any place and city can be thus discussed as an intrinsic intention to

Figure 5.3 Vernacular neighbourhood streets in the historical and restored settlement of Diriyah near Riyadh (Saudi Arabia).

Source: Photograph taken by the authors.

build a future for generations. If a critical majority develops and sustains a positive relationship and attachment to a shared place, social stability can be expected. Urbanism, the way of life in our cities, is thus a product of present socio-economic conditions, past planning decisions, and evolving political structures as well as the realm of images and intentions to build and maintain an urban habitat.

Manuel Castells (1983) identified our urban crisis as multidimensional affecting cultural, political, and economic aspects due to a new socio-spatial reality expressed in new forms of urbanism.

154 URBAN DESIGN

Figure 5.4 A contemporary neighbourhood street in Riyadh (Saudi Arabia). Source: Photograph taken by the authors.

In his paper, he introduced four main processes leading to the core transformation of urbanism in our historic period: (1) the concentration of the means of production in complex metropolitan regions challenging traditional polycentric patterns as discussed by Peter Hall (1988); (2) the on-going economic specialization of cities following the interests of capital; (3) the commodification of the city itself via real-estate markets, especially residential developments, and new modes of transportation, as discussed by Harvey (1981); and (4) the mobilization of populations leading to mass migration and new socio-cultural transformations (Figure 5.4).

In addition to the works of Manuel Castells and David Harvey, the French philosopher Henri Lefebvre had a major impact on our today's understanding of this significant period in its historical context by discussing the dialectics of socio-spatial interactions. Environments are not only experienced as pure functions linking the essential needs of our everyday life, but environments are also perceived as images, as forms representing our connections to natural orders and ourselves within them.

Henri Lefebvre introduced his methodology of studying human-spatial interrelationships via so-called rhythm analyses of everyday walks, introduced in his last publication (Lefebvre 2017). This phenomenological approach is not confined by any specific procedures but rooted in a critical discourse on images and their rhythmic repetitions. The core rhythms being distinguished are linear and cyclical rhythms. Linearity is experienced if environments are dominated by functionality and the repetition of images representing the exposure to distinctive time periods, in which the human mind is relating to the environment by assessing the dimensions of spaces and the duration required to interact with them. Thus, a constant geometric repetition of built features might enhance our spatial orientation but might also lead to a rather dominant exposure to dimensional and thus quantifiable aspects.

In contrast, a cyclical rhythm of walking experiences is not grounded in the repetition of an engineered measurement, but the exposure to natural cycles. These natural cycles represent our basic harmonies between our bodies and our environment. The exposure to natural materiality and light as well as symbols representing associations to our environmental origin and belonging can lead to spatial experiences detached from abstract and conceived measurements. These rhythms are thus grounded in forming meaningful relationships to our surroundings and each other. Henri Lefebvre therefore distinguished different forms of rhythms by establishing the theory that most of our urban experiences are polyrhythmic and contain both linear and cyclical rhythms (Lefebvre 2017).

The dominance of linear rhythms can lead to a disbalance and so-called arrhythmic experiences, which can negatively impact human health, while a stronger presence of cyclical rhythms can support 'eurythmic' experiences with meaningful harmonies. These general observations have been made by many environmental psychologists, and thus, natural environments and materials as well as the exposure to daylight are today more frequently integrated in designing healthcare environments, such as hospitals (Theodore 2016). The application in urban design practices is well understood but often less implemented in developing public spaces due to the various requirements to accommodate cars, which can be identified as a main producer of linear rhythms in our cities today.

FURTHER READINGS

Leary-Owhin, M.E. & McCarthy, J.P. (2020). *The Routledge Handbook of Henri Lefebvre, The City and Urban Society*, Routledge: London.

Lefebvre, H. (2017). *Key Writings*, Bloomsbury: London.

Schmid, C. (2022). *Henri Lefebvre and the Theory of the Production of Space*, Verso: London.

Soja, E.W. (1989). *Postmodern Geographies: The Reassertion of Space in Critical Social Theory*, Verso: London.

TODAY'S URBAN CRISIS AND THE NEED FOR SUSTAINABLE URBANISM

To explore the distinctive role of specific urban qualities required for sustainable urbanism, the general definition of a sustainable way of urban living needs to be discussed in the context of today's urban crisis. To safe natural resources, our lives must move into more efficiency regarding the quantity and frequency we consume goods and services as well as how we produce them. Henri Lefebvre is often referred to as one of the biggest critiques of modern capitalism, since he pointed out the core concern of a system being dominated by an abstract strategy rooted in a linear concept of growth and the preference of fast progress instead of a slow but equal distribution of natural resources (Wiedmann & Salama 2019b). It is however important to state that he applied the same criticism on the Soviet Union and its communist approach of top-down decision-making (Lefebvre 2009). In both systems, economies are largely planned by a few decision-makers and thus based on limited perceptions and limited periods of time. This is particularly the case of major infrastructural investments, which have become the foundations of modern urbanisation enabling both accelerated production and consumption. Few decision-makers on core investments however inevitably lead to a hierarchy of specialised producers of services and goods following the rationale of attracting increasing investment due to an already gained economic position (Fujita et al. 1999).

This phenomenon of accumulating investment in few locations and key enterprises has enabled a faster modernisation and subsequent globalisation, which can be best observed in expanding networks of airlines and their flight frequency (OECD 2010). It has however been come at a cost of increasing dependencies between

Figure 5.5 Suburban housing in Tacoma, Washington State (USA).
Source: Photograph taken by the authors.

exponentially growing consumption and environmental challenges. If any investment in a key economic sector, such as oil production, requires fast growth, an increasing investment in car production can be expected in addition to an increasing investment into required infrastructure. This cycle has affected the entire planning system of our cities and the subsequent housing supply, which has never been initiated by pure demand-driven mechanisms in any scenario of a modern urbanisation. Housing typologies can be seen as one of the biggest symptoms of our global economic development and key factors in consolidating the division of our societies rooted in either consumerism or in industrial production. The introduction of suburban housing typologies was mainly made possible by an emerging lending industry in the twentieth century. The subsequent mortgages and real-estate cycles have been fuelling consumption growth in already infrastructurally supplied region, enabling emerging service sectors and the import of industrial goods. The relocation of manufacturing industries to regions offering lower production costs included the relocation of environmental and social costs, which are the consequences of any industrialisation process (Figure 5.5) (Ahuti 2015).

Today, the urban crisis can be summarised as the result of a global economic system, which has been built on the principles of rapid growth and technological progress by trading environmental balance and social equity (Weaver 2017). On the one side, the rapid modernisation and globalisation have enabled an unprecedented population growth in most countries, and the distribution of resources however follows the rationale of prime investments in key locations and sectors resulting in a global hierarchy of economic centres. One consequence has been the phenomenon of mega cities in developing countries, which have been attracting migration due to modern infrastructure being built, sustained, and expanded in their agglomerations and the associated easier access to industrially produced goods. This has led to a highly interconnected but dependent world of cities with limited regional economic resilience (Leffel & Acuto 2018). This structure has also been the cause for an accelerated energy consumption per capita. Energy is still mainly consumed in form of fossil fuels with devastating effects on our environment worldwide. As any form of energy production is known to cause ecological challenges, the so-called renewable energy is still dependent on a variety of resources, which need to be transported and industrially manufactured. Thus, the prime objective should be ideally seen in a reduction of energy consumption per capita first.

To reduce our waste of resources and to move to a more regional and circular economy, the general idea of fast growth and associated progress is a rather complex and interdependent system to be questioned. Our postmodern condition is rooted in a continuous discourse on individualism and the end of collectively shared ideologies (van Raaij 1993). The widely shared expectation of having individual access to technologies and to spend more time for individual interests can be however seen as the result of a newly formed ideology. Major ideologies have always been the result of both applied concepts and individual experiences and postmodern ideologies are therefore no exception. This condition however needs to be discussed as a key factor within our urban crisis, which can only be overcome by a shared interest in transforming our ways of life rather than relying on top-down resolutions in form of further technological progress. The smart city idea is rooted in ideas of applying new technologies, especially digital applications, to manage data and resources more efficiently. Due to the need for significant initial investments, smart cities can however be questioned as prioritised development visions in many regions. Smart cities can support improved living conditions for various urban populations, but only if social sustainability concepts are an integral part of the overall strategy (Chen et al. 2022).

As the economist Henry George (1880) discussed in his book *Progress and Poverty*, the question of land ownership can be still identified as a key factor hindering a more ecological and socially responsible development. According to him, the privatisation of natural resources, such as land, water, and air, will lead to consistent social inequity. Many development strategies have been applied to find new forms of more collectively owned land or systems rooted in taxation. These strategies have had often limited success due to their insular approach and their uncompetitive nature versus the existing growth-driven system built on privatisation. Increasing the access to privately owned land has been very successful in accelerating economic developments and improving many individual lives in short periods of time (Sullivan et al. 2007). The downside of this approach must be however seen in a production of urbanism, which is not rooted in reducing the consumption of resources but instead being built on expanding it. Consumption itself can be increased via both population growth and the way we live. While new technologies, such as electric cars, can have a positive impact on lowering air pollution and carbon emissions in our cities, their introduction however implies a further dependency on new resources being extensively consumed (Henderson 2020).

FURTHER READINGS

Jacques, E., Neuenfeldt Júnior, A., De Paris, S., Matheus Francescatto, & Siluk, J. (2024). *Smart cities and innovative urban management: Perspectives of integrated technological solutions in urban environments, Heliyon*, 10(6).

Larco, N. & Knudson, K. (2024). *The Sustainable Urban Design Handbook*, Routledge: London.

Ritchie, A. & Thomas, R. (2009). *Sustainable Urban Design: An Environmental Approach*, Routledge: London.

Wheeler, S.M. (2023). *The Sustainable Urban Development Reader*, Routledge: London.

SUSTAINABLE URBANISM AND THE ROLE OF DIVERSITY

After reviewing this general context of our everyday urbanism and the resulting urban crisis, the main factors for a more ecological and resource-efficient way of life can be discussed. One key spatial quality promoting a more sustainable development can be summarised in diversity. This however requires a more in-depth exploration. Spatial diversity for reducing energy consumption is the result of a balance between local consumption and production dynamics and investment decisions. Any diversification of built urban environments is thus the result of local economic developments rooted in highly connected and integrated places for working and living, which however requires long-term investments. Diversification is thus highly dependent on demand-driven spatial developments, such as a big choice of housing typologies and offices for various company sizes, which require continuity. The most diverse urban structures offering various housing markets, services, and workplaces in direct proximities can be thus found in very old and consolidated cities. Their economic resilience can often be directly referred to the spatial quality of diversity enabling knowledge economies and all their innovations (Wiedmann et al. 2012).

Despite its size and deficits in more affordable housing, London's biggest quality has been its high level of spatial diversity due to its long-term development as a global centre. This has resulted in enabling one of the best transit systems with a long history, which has been partially extended and modernised in recent years, but which has been delivering in its everyday service (Rode 2017). And this effective service has only been possible due to the diversification of many urban areas and the resulting polycentricism and integration. A similar phenomenon is Tokyo in Japan and its high performance in transit-oriented development, while enabling many lifestyles of working and living in direct proximity to each other (Vazquez et al. 2023). While demand-driven dynamics and subsequent consolidation are certainly the producing factors of spatial diversity, urban planners and designers need to be highly aware of the various aspects, which need to be preserved or introduced to sustain or to enhance spatial diversification (Figure 5.6).

Planning and design for diversity therefore requires certain basics, which can be summarized as: (1) adaptive zoning; (2) old

Figure 5.6 Piccadilly Circus is a public square on top of a busy underground station (London, UK).

Source: Photograph taken by the authors.

building stock; (3) adaptive renewal; (4) differentiated grids with various levels of centrality and density; (5) integration of affordable housing; and (6) different modes of mobility. Zoning is a planning subject with a long history and various requirements. In theory, the idea of leaving land use and building requirements less regulated and open for various development applications, which can be decided case-by-case, can undoubtedly promote a bigger diversification. On the other hand, a deregulated zoning strategy can lead to rather complex conflicts, especially in the case of rapid urban development, as experienced in developing and emerging countries. These conflicts are often rooted in matters of insufficient infrastructural supply, including electricity and parking, or incompatible building uses, causing noise and air pollution next to residences. Subsequently, deregulated approaches can lead to opposite effects and middle- and higher-income groups leaving urban areas resulting in social segregation. This is then often followed by companies relocating to office parks and retail moving into shopping malls in peripheries to be more accessible by upper-income groups.

The consequence has been deteriorating city centres struggling for long-term investment for diversification.

Adaptive and flexible zoning approaches need to be thus reviewed from the perspective of the development status of urban areas. If urban areas are historically consolidated, a case-by-case approach of approving building applications can support increasing diversification. In urban areas, which are facing major growth periods attracting investment, a more regulated approach needs to be preferred to offer guarantees to a bigger diversity of individual investors, who need to count on compatible developments on adjacent sites. The tendency of deregulation to attract investments has usually been leading to more speculative dynamics and subsequent planning challenges for supplying infrastructure and reducing emerging conflicts. In summary, flexible zoning approaches can be an important key to sustain and increase diversity in already grown urban structures, such as London, and it however needs to be critically reviewed in the context of the overall development status of local economies to avoid speculative tendencies in developments. One example for this problematic effect of deregulated zoning can be found in contemporary Gulf cities, such as Manama, the capital city of the Kingdom of Bahrain. The rapid urban growth in the early 2000s led to an overemphasis of multi-story apartment blocks for international workforce on deregulated urban areas along shorelines with hardly any diversification with different typologies or functions (Figure 5.7) (Wiedmann 2013).

Another key aspect for spatial diversification is the existence and maintenance of old building stock. This aspect was already emphasised by Jane Jacobs (1993). And it is not specifically referring to heritage buildings, but buildings, which are more affordable to rent by smaller businesses, companies or in the case of housing by lower-income groups. Thus, old building stock often includes buildings which were built in the post-war era between the 1950s and 1970s, which have frequently been built on strict budgets, and which offer very limited potentials for upgrading. In many cases, this building stock has often been removed due to investment pressure in central and accessible areas. The remaining buildings are located along traffic thoroughfares or hidden side streets and can have a positive impact on enabling start-ups and smaller businesses to be part of urban marketplaces. The future integration of

Figure 5.7 Apartment blocks in Juffair (Manama, Bahrain).
Source: Photograph taken by the authors.

this building stock mainly depends on maintenance incentives and an improved public realm to increase access and urban safety. In some cases, the heritage status could be extended to certain typologies from this era, which are seen as important representations of their time (Havinga et al. 2020). This policy-driven initiative can disable their quick removal and enable a broader public debate on the values of old building stock and what it contributes to our cities in the context of diversification and sustainability. One driver of local economic development worldwide can be seen in small entrepreneurial initiatives of many lower educated groups to start businesses, from small diners to hairdressers. The replacement of old buildings would immediately limit the spaces needed for this essential socio-economic development enabling future generations to have a bigger share of evolving markets.

The next basic strategy to promoted spatial diversification is to enable more incentives and possibilities for adaptive reuse and renewal of old building stock. This strategy is directly related to the prior one, but it differs significantly by aiming for better rental return or higher property prices. While it is important to maintain and extend affordable opportunities, it is as important to attract bigger investment in renewal incentives and to integrate higher-income tenants or owners in central urban areas. This process is generally known as gentrification, and despite its danger of being a decisive factor in pushing out lower-income groups and their businesses, it can have an important role in diversification. Only bigger investments can often turn previous offices, which lost their marketplace, into other uses, such as residences. This reuse enables the building stock to be essentially renewed, sometimes completely reinvented. Heritage buildings can be saved and successfully integrated by this strategy (Arfa et al. 2022). Outdated office spaces can be restructured into smaller more adaptive co-working spaces enabling start-ups to rent spaces in best locations. Monofunctional blocks can be turned into mixed-use developments offering both residences and workplaces. This renewal is however dependent on both infrastructural supply and already established or emerging knowledge economies, built by higher-income groups preferring urban residences and workplaces. This phenomenon has been described by Richard Florida (2002) as the so-called creative class and their distinctive impact on urbanism.

The basic strategies of restricting the removal or change of old building stock as well as enabling easier pathways for renewal and reuse by less strict heritage policies can be contradictory if they are not well balanced. It is thus a major responsibility of local authorities to guide gentrification processes while protecting lower-income businesses and residents by enabling them to benefit from more local economic activities. In addition to policies, one key planning and design aspect in this regard is the thorough understanding of districts and their street grids. Some street segments have higher levels of centrality and are thus more accessible by visitors and upgrading or densification processes in these areas are usually a direct result of more investment interests in these locations. Side streets, on the other hand, are less directly connected and offer valuable spaces for smaller businesses and the integration of residential use. The grid structure must be therefore carefully investigated and understood. Vernacular settlements worldwide have been built by market activities and thus daily walking routines instead of top-down ad hoc planning to rationalise urban systems via infrastructural constraints. All historically and organically grown settlements show a rather clear differentiation between highly connected market streets and less connected secondary streets leading to more quiet and private areas. The transition speed from more accessible to less accessible spaces can however heavily vary depending on culture and climate (Hillier 1996).

The more regular urban grids are the less differentiation they can offer and the more often they can suffer from less diversification in each urban block. Thus, cities, such as Glasgow, Barcelona, or many cities in the United States with regular grid structures, often suffer from a spatial segregation tendency of higher and lower-income markets. In contrast, cities, such as London and Paris, offer a big variety of integration levels due to an irregular grid and a variety of block sizes and densities. This irregularity might reduce certain aspects of efficiency and orientation, but it is a major factor in enabling a bigger diversity of rental prices in direct proximities and thus enabling a bigger local integration of different markets. So, major headquarters of transnational firms in London can be directly surrounded by smaller companies located in adjacent blocks along small side streets. This heterogeneous network of location advantages can be seen as a major precondition for complex knowledge economies and creative industries depending on innovations by smaller and more flexible service providers.

Another strategy for spatial diversity is a proactive integration of affordable housing strategies to counteract gentrification trends. Social housing supply has a long history rooted in the first industrialisation periods and the accommodation of low-income labour. The top-down strategies to develop social housing have often led to rather large areas being occupied by one housing typology with adverse effects on adjacent areas due to the move of higher-income groups. This move has often been caused due to social infrastructural matters, such as schools, or safety concerns. Therefore, it is important to enable a more careful spatial distribution of social housing schemes. To achieve the housing supply, social housing driven by government initiatives has often been replaced by policies on new developments securing a certain percentage of housing units being built as affordable options. Other approaches in developing affordable housing can be the introduction and support of cooperatives or co-housing for middle-income households, and so-called housing associations for lower-income households. Housing associations are privately owned but structured as non-profit organizations. Any successful implementation of affordable housing strategies however depends on land price matters. Thus, a strategic coordination between new opportunity sites and the integration of affordable housing shares, like infrastructural matters, which often must be co-developed by private sector incentives.

Finally, spatial diversity enabling a high diversity of social activities in short distances can be promoted by a better integration of different mobility choices. From a sustainable urban design perspective, walking needs to be achieved for all everyday destinations. While workplaces can be seen as the most challenging to be integrated, walks to schools, grocery stores, and other basic social services as well as leisure spaces can be achieved via modern planning and catchment areas. Highly activated pedestrian connections can help the overall diversification of districts due to synergies, such as daily walks to high streets and a direct exposure to different shops and businesses. These pedestrian connections must be well designed and maintained, and ideally cycling routes are equally integrated along main corridors. To connect all key districts and main urban centres, public transit networks need to be developed in strategic locations to offer a good alternative to daily commuting by car. Car sharing is another important strategy to reduce individual car traffic and to enable access to a more flexible form of mobility needed in certain occasions. In addition, car sharing can often be better combined with a more effective electrification by avoiding the overuse of local grids (Schlüter & Weyer 2019).

In conclusion, spatial diversity is produced by our local economies interacting with investment trends affecting land prices, infrastructural costs, and building stock. These investments need to be either more guided or less regulated to sustain or increase spatial diversity. If these investments are more in sync with demand-driven dynamics in local production and consumption dynamics, we can generally observe a higher diversity than in scenarios, in which local economies are underdeveloped. In cases, in which bigger developments, such as entire new town projects, are being built, diversification processes cannot be expected until local economies are forming over time. This reality of people-dependent spatial diversification over longer periods of time can be seen as one of the main factors, why many large-scale developments can often fail in delivering sustainable forms of urbanism. If cities are created by few market makers, such as governments or bigger private sector funds, they can only express limited markets, such as the construction industry and all its related service sectors. Diversification can however only happen in functioning economic systems with more local entrepreneurial incentives in various scales, and this can only be achieved via both time and proactive urban management enabling all key basics to be considered and implemented.

FURTHER READINGS

Glaeser, E. (2012). *The Triumph of the City*, Pan Macmillan: London.

Hillier, B. & Hanson, J. (1984). *The Social Logic of Space*, Cambridge University Press: Cambridge.

Kropf, K. (2017). *The Handbook of Urban Morphology*, Wiley: Hoboken, NJ.

Oliveira, V. (2018). *Urban Morphology: An Introduction to the Study of the Physical Form of Cities,* Springer: London.

SUSTAINABLE URBANISM AND THE ROLE OF BELONGING

After reviewing the basics required for spatial diversification produced by forming local economies, the image production of cities needs to be seen as the second pillar of enabling a more sustainable form of urbanism. Images of places and their imageability are often discussed in the context of spatial practices, and especially along the theoretical crossroads of human geography and environmental psychology. Images can cause a sense of belonging, often discussed as place attachment, which is an important precondition for people being invested and becoming a part of a specific neighbourhood, districts, and city. While marketplaces are present-moment conditions produced by investments and the everyday interplay between production and consumption, the experiences and associated images of living in a certain place create a future development path rooted in a sense of belonging. This can be best explained by reflecting quality-of-life aspects of any city. A shrinking quality of life will be first experienced before it will lead to negative consequences in future, such as the move of higher-income groups and long-term investments. Everyday experiences can be translated into sequences of images and impressions. Repetitions are frequently detected and are most representative of a place, as discussed in the context of Henri Lefebvre's rhythm analysis.

The multitude of these everyday representations of our lives and their associations can lead to a rather big complexity, since they are not the same from individual to individual. On the one side, we are sharing the same functions of our physical senses, but on the other side, our socio-cultural backgrounds, our past life experiences, and our concrete individual life stages can lead to a variety of experiences and their interpretations including their importance to us (Schmid 2009). It is therefore useful to distinguish different categories of images by acknowledging that spatial configurations and their sequences can be interpreted as a communication between places and us. The main categories can be introduced as following by referring to other theorists (Morris 1971; Foucault, 1972; Hoshino 1987) with images triggering: (1) stimulation and excitement; (2) safety and well-being; (3) aesthetics; and (4) cultural values.

Stimulating and surprising experiences are causing our immediate interest, and they can cause this effect without being simultaneously representing a language of safety, aesthetics, or shared cultural values. In many ways, postmodernism can be best introduced as a design language often attempting stimulation in spaces without being rooted in historical values or an over-emphasis of aesthetic harmonies. Our recent decades of extensive consumerism have been combined with new urban landscapes representing images of stimulating individuals to visit certain places for new consumption experiences. From very plain and direct large-scale advertising (e.g. Piccadilly Circus, London) to entire landscapes of lights and excessive built configurations (e.g. Manhattan, New York), stimulation via built environments has often proven to be a very impactful way of attracting attention and creating places as recognizable brands. The consequence has been entire cities being envisioned, planned, and built on that premise. The best examples can be found in Las Vegas in the United States, and Dubai in the United Arab Emirates. This phenomenon is sometimes referred to as event urbanism (de Oliveira 2020). One specific stimulation can be found in images produced by high rises, often used in urban politics to underline certain visions (Gassner 2019). The archetype of a tower has been symbolizing the idea of uniting and leaving earth towards the sky across all cultures and is thus resonating with many spectators. This often positively experienced stimulation via extreme forms and colour combinations can be found and discussed in the context of an evident need for escaping from everyday lives, which have been dominated by linear schedules and repetitions (Figures 5.8 and 5.9) (Cohen and Taylor 1993).

If stimulation can be caused by certain spaces and their representations, another set of images can be found in representations signalling safety and well-being. Visual stimulations can often

Figure 5.8 The pedestrian experience in Dubai Marina (Dubai, UAE).
Source: Photograph taken by the authors.

Figure 5.9 A scenic walk exposed to nature and the historic townscape (Salzburg, Austria).

Source: Photograph taken by the authors.

only last short amounts of time before repetitions lead to decreasing attention. Images representing safety and wellbeing on the other hand are long-lasting impressions of places and still very rooted in our daily experiences without being abstract concepts. As stimulations can be found in the extraordinary and unexpected signals (e.g. via unusual dimensions, geometries, or colours), safety is often found in images representing reliable and predictable environments nurturing and protecting us. One visual input can be experienced in images supporting our direct orientation, while at the same time avoiding a repetitive and simultaneous over-exposure to too many places. Thus, big road axes in our cities ease orientation but can also lead to an overwhelming and exhausting experience of long distances. Representations of safety and well-being can also be found in shade offering a more comfortable walking experience or physical elements protecting us from traffic (e.g. safe crossings) and its adverse effects. Another set of representations nurturing our impressions of safety can be found in socially active streets and places. Therefore, it is often recommended to accommodate more active frontages, to enable space for more social interaction, and to have sufficient street lighting.

Comfort, safety, and well-being are very direct expressions of our built environment and frequently assessed during any everyday walking experience, including the materiality and cohesive surface of sidewalks and opportunities to rest. The latter is particularly important for children and disabled and aged populations. A stimulating environment does not necessarily need to be comforting, and the same can be found vice versa. It is thus a matter of integrating elements producing these impressions in an appropriate balance. The general need for safety has often led to more walled residential compounds in developing and emerging countries, which again have contributed to street sections with less active frontages and long stretches of urban walks along big blocks without any market activities or other forms of social interaction. Thus, individual safety matters can often produce images in public spaces associated with a lack of shared urban safety and a declining neighbourhood life.

Beyond these very direct expressions of our built environment, which we read via our senses, past experiences, and direct interpretations, we also need to differentiate representations causing aesthetic moments and connections to people and their culture. The field of aesthetics and the role of harmonies and disharmonies can be seen as rather wide and abstract. Some thinkers have tried to find links between music and spatial sequences (Reynolds & Reynolds 2022). Others have tried to find links between the individual positions on aesthetics in the context of an existing but continuously changing "Zeitgeist" (Graham 2006). In the context of Henri Lefebvre's theory, the main idea of harmonies in our spatial experiences can be found in the matter of an increasing exposure to cyclical rhythms and thus impressions of timelessness. One shared aesthetic can be thus found in nature itself, and many studies have shown positive effects on mental health in natural environments (Ewert et al. 2019). This realm of representations linked to mental concepts of harmonies can include matters of geometric harmonies (e.g. golden section), responsive and compatible colour palettes and human dimensions, as well as the direct exposure to natural materials and their expressions. Beyond these rather general aesthetic associations, we need to distinguish representations connected to specific local places and their communities.

Every culture has its very distinctive language via spatial forms, often introduced as so-called vernacular spaces. These vernacular expressions are rooted in regional materials, craftsmanship, as well as specific spatial practices, such as processions or seasonal festivals. Heritage architecture is a direct expression of many aspects of local vernacular languages and their historic evolution. Like a collective memory, historical built substances are linking us to past generations and their experiences. The subjective value of these images can cause a rather significant sense of belonging, which is however often differently perceived and shared by newly migrated populations or visitors. The external view on cultural representations can often be interwoven with aspects of stimulation rather than impressions of familiarity. Urban tourism highly depends on cultural heritage, while many tourists and their new spatial practices and needs (e.g. hotels) can threaten the original and authentic experiences (Xue 2022).

In conclusion, our everyday form of urbanism can be highly impacted by spatial languages and their narratives, collectively produced and experienced. The various lived sequences of images in our everyday urban environments can stimulate, comfort, and aesthetically please, as well as culturally connect us. There can be however opposite experiences in form of sequences of images boring, threatening, disturbing, or alienating us. In most urban spaces, a big variety of impressions can be collected, and we can generally observe and state that most positive interpretations would contribute to our personal sense of belonging to a specific place. This sense of belonging is a key factor for both more social interaction in public spaces and more shared ownership preventing vandalism and other adverse effects of place abandonment. Placemaking for place attachment is thus to be seen as core objective of any urban design intervention. It is therefore important to theoretically review what impressions are currently existing and what can be introduced by respecting and integrating local narratives.

FURTHER READINGS

Chase, J., Crawford, M. & John, K. (2008). *Everyday Urbanism: Expanded*, The Monacelli Press: New York.

Mossabir, R., Milligan, C. & Froggatt, K. (2021). Therapeutic Landscape Experiences of Everyday Geographies within the Wider Community: A Scoping Review, *Social Science & Medicine*, 279.

Schmid, H. (2009). *Economy of Fascination: Dubai and Las Vegas as Themed Urban Landscapes*, Borntraeger: Stuttgart.

Skelton, T. (2017). *Everyday Geographies*, Wiley: London.

SUMMARY

Both diversity and a shared belonging to places are the two essential pillars of producing urban spaces with the best potentials for resource efficiency. One key to this efficiency are short distances between living, working, and everyday leisure activities. A high level of spatial diversity would allow the integration of many different housing markets resulting in mixed densities, commercial developments, green spaces, as well as various types of social infrastructure and modes of mobility. It is however important to note that this spatial diversity is usually the result of many decades of local economic development, and it is people and their markets who usually have the biggest impact. Urban planners and designers have the distinctive role in identifying local markets and their potentials to unlock their positive dynamics in new developments by preserving old building substances, by enabling mixed land uses and densities, and by connecting places in strategic ways. An even more complex challenge is the integration of resonating images and their narratives to support a local place attachment, which is a key factor in promoting more social interaction in shared public spaces leading to neighbourhood and district formations.

As discussed in previous chapters, neighbourhoods should be seen as socio-spatial entities, and they can only be formed by everyday community interactions. These interactions however depend on shared public spaces and their frequent visits, especially in the case of neighbourhood streets and shared local marketplaces. The resulting experiences can promote a shared sense of belonging and thus many synergies in maintaining and improving the surrounding built environment. This collective engagement eventually promotes long-term investments, which are key for improving the quality and energy performance of buildings as well as better mobility choices. Neighbourhoods are the cells of our cities and thus the main determining factors of our way we live. If neighbourhoods provide a wide range of activities, daily commuting can decrease. If neighbourhoods are built on frequent social interaction, many new synergies, such as car sharing, can be enabled to further reduce the waste of resources. This people-driven transformation towards a more sustainable form of urbanism is essentially built on a combination of accessing local resources and more time. Local resources can be divided into shared physical resources, such as land, air, and water, as well as human resources, which include the potential creativity and collaborative synergies. These synergies are however often demanding certain levels of education as well as social and economic resilience. Growth-driven dynamics have however led to place hierarchies and unhealthy competitions between places and cities leading to a disbalanced distribution of both resources and time.

This chapter therefore needs to conclude by reconnecting to the theory of Henri Lefebvre and his idea of differential spaces, and an urban revolution led by collaboration rather than any top-down intervention or plan. By recognizing the potentials of shared resources and time, our consumption- and growth-oriented forms of urbanism, which have been leading to both fragmentation, segregation, and social isolation, can be replaced with new development models. Initiatives, such as cooperative housing models, have already shown potentials to integrate our everyday interests for better and more resource and time-efficient ways of urban living. Moreover, several studies have shown the vast potentials of urban farming if knowledge and access to green spaces can improve (e.g. Pradhan et al. 2024). The strategy of digitalising infrastructural services and other urban management matters can become another valid contribution for saving resources, such as water and electricity (Lange et al. 2020).

Any new technology is however again bound on certain resources, and the dependency on technologies will always lead to a conditioned rather than self-defined form of urbanism. Resource efficiency however needs to become a shared value and consequently a shared effort in our everyday lives. The continuous and repeated cycles of trusting hierarchical and technology-based solutions have shown its major limitations and will not end but sustain social inequity. This inequity however means unequal access to changing development dynamics via co-ownership and participation. In the next chapter, urban design basics will be therefore explored from the perspective of the specific role of urban designers and their services as integral part of urban governance.

6

URBAN DESIGN AND GOVERNANCE

After reviewing and discussing the key theories of urbanism and urban design history, this chapter will provide an overview of how urban design is managed as part of urban governance and how five main project types can be distinguished. In the context of our current digitalisation, urban design as any service industry can be aided by new applications supported by machine learning. The scale of this impact however highly depends on if cities and their development paths continue to be defined by new technology, as in the case of smart city strategies, or if they are going to being soon impacted by new market realities and the need for more participatory decision-making. The assumption that technology-led development paths are superior to achieve sustainable urbanism has already been questioned in Chapter 5, mainly based on the rationale of social equity concerns in most places.

The most important aspect of any machine learning is to provide more informed solutions due to growing data sets and increasingly dynamic algorithms to make centralised decisions for more efficiency (Chamorro-Premuzic 2023). This, however, can lead to a standardisation of urbanism, the way we live in our cities with less co-ownership in marketplaces, and thus less bottom-up demand-driven dynamics. While key functional aspects might be assessed by new applications following available empirical metrics and increasingly convincing digital twins providing more big data, the qualitative aspects of form, aesthetics, and meaning of places can only be broken down via abstraction rather than any shared consent. The form and language of spaces will remain produced by the future course of human attachments, such as shared values and the

DOI: 10.4324/9781003251200-6

idea of the "Zeitgeist" and *genius loci* interpretations (Eisenman & Iturbe 2020). New automated systems might be able to follow adaptive paths depending on determined frames which can neither fully integrate the complex human condition, whose present moment is built on the grounds of both past and future ideas as well as intentions.

As pointed out in the previous chapters, we experience spaces as sequences of images in combination with other input from our senses, and each image thus contains several layers of different information. While information can be gathered by AI applications, the process of collecting this information is in itself a process of reduction and abstraction. The entire phenomenon of outsourcing both the assessment and solution-making for living places to machine solutions will be rather quickly outlived due to a completely one-sided and thus a very fragile approach by rationalising spaces as pure mathematical functions without acknowledging their equally important dimension of timeless forms expressing still evolving shared languages and thus human relationships. It will therefore remain the domain of urban designers to read spatial conditions holistically, to communicate the various key findings via strategic prioritisation, and to connect to all main stakeholders and thus shared markets enabling new spatial conditions for a big range of dynamic interests.

Any place an urban designer is engaging in is the subject of human interests, and those interests require a translation into functioning as well as resonating placemaking efforts. In general, we can observe that the more people are locally involved, the more shared ideas can be identified and integrated. In the following, this chapter will identify and discuss key aspects of governance and urban design services and their core challenges in implementing people-based solutions for sustainable urbanism. First, urban governance itself needs to be assessed regarding its current organisation and challenges before distinctive project types can be introduced and discussed. Since this book was not written for any specific national context, governance can only be explored from a general global perspective by reflecting on recent economic developments and the role of cities by however distinguishing developed and developing regions.

URBAN GOVERNANCE

To introduce urban governance, governance itself needs to be discussed first as the organisation of decision-making regarding both new development visions and implementing strategies including ways of data collection (Song et al. 2023). The governance of a city will always regard the city as a continuous project, a project which is facing challenges while having various distinctive development potentials. Based on the realities of land ownership and investment dependencies, governance has become an increasingly complex public-private operation, in which top-down decision-making has become a key factor for efficient solutions in key domains, such as infrastructure, land uses, and building regulations.

The increasing costs of infrastructural supply can often only be met by enabling more private sector engagement leading to a cycle of investments, which again needs to be managed via regulations how and what development can be added. Urban agglomerations are thus the product of a continuous attempt for modernisation via new infrastructure attracting investments requiring more infrastructure and future investments. Governance as the organisation of decision-making cannot be reduced to government authorities only, despite their decisive role in implementing policies and development strategies, such as urban and infrastructural planning. Major private sector stakeholders in investing in the development of the city are as important since their marketplace can only function and expand if the city itself is functioning and growing. This urban growth does not necessarily imply physical growth itself, but growth in the regional and global network and hierarchy of cities.

To understand the reality of the hierarchy of cities, it is important to revisit John Friedmann (1986) and Saskia Sassen (2001), both outlining that cities have become important hosts of control centres of our global economy. Peter Taylor and Ben Derudder (2015) have furthermore investigated the complex interplay of headquarters within knowledge economies and their strategic location choices. These choices highly depend on geopolitical aspects as well as matters of inner social, economic, and political stability, but they also depend on growth driven by infrastructural supply via ports and airports to support a high level of global connectivity. The main financial centres, such as London, New York, Singapore, and Hong Kong, have become important nodes of directing urbanisation processes worldwide due to the scale and frequency of various investment flows. If cities are identified as potentially important factors in accelerating the growth of global production and consumption, investments in new infrastructure including ICT can permit new development dynamics (Pradhan et al. 2021).

In the Global South, the late modern urbanisation and associated rural–urban migration in many regions have led to early dependencies on already established networks, leading to major difficulties for local governance (Randolph & Storper 2023). The still existing divide between developed and developing regions has led to persistent economic disparities, which need to be seen as the most important context for any urban governance to manage local developments (Hickel 2017). Only in the case of a modernisation of infrastructure, improved connectivity can support the location choice of both investments and companies to enable developments. While the macro infrastructural concerns are often decided on global, national, and regional levels, one local reality of any existing settlement remains: It is the everyday habitat of numerous existing and future communities, and it is essentially human resources, who are decisive for any development enabling governance to manage growth in new directions (Das 2022).

Growth management can therefore be seen as one of the most important objectives of any urban governance in our global economic context. On the one hand, the stimulation of growth with infrastructural incentives is an important factor for modernisation and the improvements of the general quality of urban life, such as the access to electricity, water, and, in recent decades, internet. On the other hand, rapid growth is only possible via extensive migration processes, which are very difficult to manage for any local governance due to the missing opportunities for demand-driven markets and the dependency on supplying housing and infrastructure first (Wiedmann & Salama, 2019). Cities are therefore often perceived as growth machines enabling the access to technological progress and simultaneously cities have been playing an important role in sustaining social inequalities despite their provision of modern lives (Molotch 1976; Harvey 1975).

The top-down structure of urban governance can be best assessed by beginning with reflecting these global economic conditions. Accelerated economic growth is necessary to enable rapid modernisation with the ideal perspective of future consolidation, which is less dependent on physical growth. This transition is however widely understood as difficult to implement and manage (Nelson & Duncan 1995). Fast initial growth is however only possible with entering global networks of trading and investing. Thus, these decisions on infrastructural access impacting urban agglomerations are made on national and supra-national levels and are setting the context for any local urban governance. It can be thus argued, as Jane Jacobs (1970) pointed out in her book *The Economy of Cities*, that the economic decision-making on macroscales is following the rationale of key urban agglomerations and their growth requirements rather than the far more complex requirements of entire regions and nations. Thus, cities have also historically become the shapers of regions rather than following the theory of an organic evolution, in which regions produce villages and big villages transform into cities. It is the accumulating knowledge in trading networks which leads to the emergence of bigger cities which then shape their surrounding regions and finally global economic conditions.

The next level of decision-making can thus be found in the actual management of urban agglomerations, which can be organised in various ways. In many cases, ministries can be still heavily involved in managerial affairs, such as major infrastructural development and policies, since macro-economic interests are mainly addressed to governments, who are facing challenging realities of either opening or controlling growth incentives. The day-to-day management of our cities can be subsequently found in the interlink and transition from national governments to municipalities and their local authorities (Raco 2009). These municipalities are mainly responsible to introduce rather general development visions while managing the everyday requirements of their urban agglomerations. Urban planning authorities can be important stakeholders of urban governance. Their decisive role is however based on their capacity to monitor and assess current demographic and spatial conditions as well as to implement responsive strategies and policies rather than to keep outdated ones.

Local growth management is thus first rooted in the comprehensive collection and surveillance of data. The recent advancements in satellite technologies and the overall digitalisation have enabled governance to collect more big data in real time, which is a key factor in managing increasingly complex urban agglomerations and their infrastructural requirements (Batty 2016). Spatial conditions can often be rather quickly surveyed and stored in geographical information systems, while social and economic conditions are far more challenging to monitor and to keep updated. By approaching cities as growth machines in global networks, the top-down view on abstract and only periodically updated census data can often lead to a distanced view on communities (da Cruz et al. 2018). It is however empirically known that it is the economic resourcefulness and long-term commitment of those communities enabling a more integrated development including consolidation.

The here-often used terminology of consolidation requires some further definition and discussion. Consolidation is defined as the process of combining several actions into a single more effective and coherent whole. In the urban development context, this means that dependencies on urban growth and associated urban sprawl need to be lowered, and existing urban areas need to be transformed towards a more effective land use and balanced densities (Shaw & Houghton 1991). This process is highly dependent on the existing land ownership structure and the economic context of local communities, who can sometimes resist consolidation initiatives by introducing higher densities in residential areas (Ruming 2014). After World War II, governance worldwide has been fixated on rapidly modernising urban lives by enabling both better infrastructure access and more consumerism, by however ignoring the consequences of a new form of everyday life rooted in daily commuting rather than active neighbourhoods forming urban communities (Figure 6.1).

While government institutions, from ministries and their macroeconomic development strategies to regional and local municipalities and their planning and development authorities, have a decisive impact on the structure of governance, the role of the private sector needs to be discussed as an increasingly integral part of decision-making processes (Cook 2009). The complex risk-taking of major development projects requires an increasing amount of expertise,

Figure 6.1 A typical UK city centre street showing the decline of retail activities (Nottingham, UK).

Source: Photograph taken by the authors.

which often cannot be found in government institutions and instead accumulate in entrepreneurial environments. Thus, major developers of infrastructure and real estate often form public–private partnerships with joint decision-making including the negotiation of policies and the prioritisation of projects.

In many regions, it can be stated that the main reasons for a more privatised form of urban governance have been the combination of limited capacities in governmental institutions and the increasingly challenging market environments requesting faster and more integrated decision-making to mitigate risks. This on the other side has accelerated the often-criticised commodification of cities and the overall understanding of cities as prime investment opportunities for private and often international investors rather than conceiving cities first as shared marketplaces enabling thriving communities, resulting in a vibrant urban culture. Today, it can be even argued that the images of local city culture have become an important part of strategies attracting investment, especially due to international media attention (Ramadhani & Indradjati 2023) and expanding tourism (Soltani et al. 2017). This phenomenon is often discussed as city or place branding (Moilanen & Rainisto 2009) or the economy of fascination (Schmid 2009).

In summary, governance is the organisation of decision-making, and in a capitalist context, the prime decision-making is led by strategies for initial investment and development. Most major investment however expects quick return via growth incentives, and the challenges of managing physical growth of cities can only be tackled by initiating diversification and consolidation processes. This integrated development can however only happen if cities are less dependent on physical growth by playing more active and critical roles in global hierarchies of advanced knowledge economies. This again requires an increasing or at least stable quality of urban life to attract the most educated and skilled human resources (May & Perry 2018) and to avoid their loss due to international migration, often discussed as "brain-drain" (Netek et al. 2022).

Today, cities are still conceived as important growth catalysts for national economies, while they are increasingly difficult to manage due to their expanding size, problematic boundaries, challenging socio-economic realities, and the on-going spatial fragmentation instead of polycentric and integrated development. As already outlined in previous chapters, the support of a more sustainable form of urbanism cannot be compromised by short sighted growth strategies. This will however depend on new forms of governance enabling more bottom-up incentives to improve the quality of life in neighbourhoods forming districts and districts forming cities. Main city centres are equally challenged to reinvent themselves as more integrated urban habitats rather than as just mono-functional commercial hubs due to the shrinking need for traditional office space and retail (Wiedmann et al. 2022).

FURTHER READINGS

Florida, R. (2002). *The Rise of the Creative Class*, Basic Books: New York.
Hodson, M., Kasmire, J., McMeekin, A., Stehlin, J.G. & Ward, K. (2020). *Urban Platforms and the Future City Transformations in Infrastructure, Governance, Knowledge, and Everyday Life*, Routledge: London.

Sassen, S. (2018). *Cities in a World Economy*, SAGE Publications: London.
Taylor, P. and Derudder, B. (2015). *World City Network: A Global Urban Analysis*, Routledge: London.

THE ROLE OF URBAN DESIGN

One of the biggest threats for a long-term decline of a shared quality of urban life can be found in the urban development strategies invented in Western cities in the first half of the twentieth century. To initiate investment incentives and growth via a more predictable expansion of consumption and subsequent service sectors, land in peripheries was excessively added to our cities as new dormitory suburbs (Hall 1988). The first time in history, consumption and production were spatially divided on a macro scale. This was accelerated by the second Industrial Revolution and the car as new mode of transport, which has emerged in combination with individual home ownership as another key economic factor in increasing individual consumption. In many respects, the so-called crisis of cities can be directly referred to a new form of urbanism rooted in cars as prime mode of transportation and all its consequences for the individual everyday urban life (Castells 1978; Jacobs 1993).

While cars first enabled the supply of private land ownership on a new scale, which was perceived by many Western societies as a prime concern of individual safety and independence after the negative experiences of urban living during industrial revolutions, cars themselves can be identified as the most critical concern regarding the long-term quality of urban life in most cities. First, cars have led to a third and fourth hierarchy of our street networks: the various highways interlinking cities and major urban thoroughfares interlinking all major districts and suburbs. This infrastructure is not only expensive to build and maintain, but it has also led to a spatial divide of districts and their neighbourhoods due to less frequent crossings for pedestrians and cyclists. These major road networks have also led to an increase of air and noise pollution in adjacent areas and an entirely fragmented experience of our town- and landscapes. Due to continuous urban growth, new lanes must often be added to lower the increasing traffic congestion. This congestion needs to be identified as another major factor in lowering the quality of life due to the individual time and thus economic loss.

Increasing energy costs in recent years have further exacerbated this individual loss. Thus, it can be observed that the individual gain of individual land ownership in remote suburbs has often lost its appeal versus the overall loss of time and money in many cities. Subsequently, higher-income groups have often returned living in cities or in proximity, often leading to gentrification (Hochstenbach & Musterd 2017).

Mega cities in the developing world have been challenged in establishing quality of urban life mainly due to rapid growth via rural–urban migration contesting basic infrastructural supply. The initiated modernisation and industrialisation of these cities has led to hardly manageable growth dynamics resulting in informal settlements or mass housing with limited qualities. Another directly associated consequence has been the emergence of gated communities for middle- and higher-income households, who again have turned to cars as the main mode of transportation. The subsequent social segregation in combination with a generally underdeveloped road network, and high urban densities in central places as well as continuous urban sprawl have led to even more severe traffic congestions in developing regions as in many cities in North America, Europe, or Australia. Overall, it can be stated that establishing a high quality of urban life is decisively dependent on strategic growth management, since rapid growth usually leads to disintegration and fragmentation on socio-spatial as well as functional levels following the rationale of quick investment returns. The resulting fragmented urban landscape is a major challenge for any urban governance, which often finds itself in prioritising immediate measures to reduce evident conflicts rather than being in a financial position permitting long-term solutions. The best example is the expansion of existing road networks rather than the establishment of a more efficient public transport.

Growth management via transit-oriented development can however only be effective if the new public transit networks connect and link established centres on district and neighbourhood levels. The question of establishing shared centralities is the question in which urban planning is often meeting urban design matters. As discussed in previous chapters, urban design is following the general objective to configure the function and form of our shared urban spaces. The largest percentage of public spaces can be found

Figure 6.2 Cars parking at a new metro station in Riyadh (Saudi Arabia).
Source: Photograph taken by the authors.

in all streets connecting neighbourhoods, districts, and city centres. It can be argued that all other open public spaces, such as plazas and parks, are part of the overall movement network, and in many instances, they are nodes or destinations. Any street must be thus understood as both a connection and a shared centrality within different hierarchies (Figure 6.2).

A basic neighbourhood street is connecting all households along one shared corridor, and before cars entered our lives, the neighbourhood street was a much more frequently visited space with daily social interactions. These neighbourhood streets usually connect shared marketplaces and social infrastructure forming the next level of centralities before all neighbourhoods are connected to a major district centre, traditionally formed by a larger high-street or market square supplying a variety of commercial services and cultural venues. While planning is mainly concerned with infrastructural supply in combination with interdependent land-use and density concerns, urban design is needed to enhance the actual spatial experiences by enabling comfort and safety as well as an aesthetic and culturally responsive built environment.

The obvious divide between planning concerns being focused on land management via individual transportation solutions and urban design being focused on more connected and well-designed public spaces has significantly been overcome in scenarios of consolidation, especially in cases of economic stability and the shared aim of lowering traffic and all its negative effects. In these scenarios, urban design can play an accelerating role in further integration and levelling-up processes by promoting more social interaction and engagement at a local level thereby diversifying local marketplaces and enabling a well-serviced public transit connecting all key centres. Today, urban designers are often part of governance by being employed in local planning authorities or consultancies to survey, assess, and decide strategies for upgrading public spaces to support transit-oriented development, vital city centres, and more active neighbourhoods. Urban design is a critical component in the case of larger development projects with major decision-making on centralities and their overall connectivity and shared experience in form of new or modified townscapes. It is thus a critical design service provided by trained architects and planners for both private sector stakeholders, such as master developers, and city councils or municipalities requiring external support.

In summary, urban design needs to be outlined as one of the central disciplines to support the quality of urban life by lowering car dependency and by promoting active and resilient urban spaces. Only by assessing the actual everyday experience of urban habitats, decisive changes in spatial configurations can be made to return to a more sustainable development and to overcome divided and fragmented urban landscapes. Since it is well-known that street networks, which were formed before cars were introduced, are significantly better integrated, the modern grid development needs to be questioned worldwide. While accelerated growth patterns are significant risks in establishing quality public realm, strategic measures must be early reflected to enable a future integration of more shared spaces and shared forms of mobility. One of the biggest key concerns is the overall block size and the hierarchy of streets enabling a better distribution of traffic and potentially traffic-free connections for pedestrians. This can lead to the effective establishment of other modes of transportation and a potentially greater mix of land uses and densities catering for a variety of income groups and their economies.

FURTHER READINGS

Black, P., Martin, M., Phillips, R. & Sonbli, T. (2024). *Applied Urban Design: A Contextually Responsive Approach*, Routledge: London.

Greenwood, S., Singer, L. & Willis, W. (2021). *Collaborative Governance: Principles, Processes, and Practical Tools*. Routledge: London.

Medrano, L., Recamán, L. & Avermaete, T. (2021). *The New Urban Condition: Criticism and Theory from Architecture and Urbanism*, Routledge: London.

Prominski, M. & Seggern, H. (2019). *Design Research for Urban Landscapes: Theories and Methods*, Routledge: London.

PROJECT TYPES

Since urban design can be generally understood as the decision-making of spatial configurations regarding shared urban spaces, no urban designer can operate in isolation from all the various and sometimes opposing interests of communities, markets, and policymakers. The definition or redefinition of public spaces is thus a highly communicative endeavour due to the task of integrating all stakeholder interests and of finding the best consensus. To provide an overview of key urban design services, it is important to distinguish five main project types and their specific contexts: (1) new town projects; (2) greenfield settlements; (3) redevelopment of underused and brownfield sites; (4) urban renewal and conservation; and (5) design compendiums and policies. Some project types are more common in developing and rapidly growing regions, while others are more common in already consolidated and developed urban agglomerations.

New town projects have a specific history to manage urbanisation processes and distribute urban growth. One example is Milton Keynes in the United Kingdom, which was founded 50 miles (80 km) north-west of London in the 1960s as an essential part of a housing strategy and revisiting the ideals of a garden city (Wray 2016). New town projects were very common projects in Western countries during the first decades of car-based urbanism and the general modernist vision of a more merged urban and rural landscape. This has however led to a legacy of these new towns to be rather car dependent and often functioning as satellite cities with limited local economy. In the case of Milton Keynes, it can be argued that

Figure 6.3 The commercial centre of Milton Keynes New Town (UK).
Source: Photograph taken by the authors.

the role of urban design was rather reduced to a small-town centre with its distinctive architecture and shared public realm (Edwards 2001). In recent decades, new town projects can mainly be found in the developing world, where the initiation of entire new cities via master planning has in itself its modern history (Figure 6.3).

After gaining national independence during the second half of the twentieth century, entire capital cities were planned and designed by modernist visions. Examples can be found in the case of Brasilia in Brazil, Abuja in Nigeria, or Riyadh in Saudi Arabia. All these cities share that car-based urbanism has been favoured to enable more access to private land ownership as a key driver of local economic development. This has led to the role of urban design defining entire new townscapes, which have often become dominated by large-scale landmarks and big open spaces to express a unified and modernist vision of a city rather than enabling a diverse differentiation of public spaces rooted in communities and their cultural distinctiveness. New town master planning can be thus regarded as a rather challenging task for any urban designer since the main clients are primarily seeking an ad hoc imageability of a new place to attract attention and subsequent investments.

Today, this image creation of a new town project often needs to follow global trends and must propose a combination of advanced infrastructures, green landscapes, and ideally a suggestion of unlimited scale. The consequence has been an evolving and repeated master planning approach of combining a first main grid with certain land-use flexibilities and large green open spaces, often in combination with high-rise agglomerations to mark an emerging centre. Examples can be found in many Chinese new town projects, such as Lingang in the region of Shanghai, or in the Middle East (e.g. Lusail City in Qatar). Many projects share being defined by a distinctive geometry, which can be easily identified, even from satellite images. Forest City, Johor in Malaysia is a very specific case of a recent new town project, which has been initiated by private sector interests and which is currently known as one the largest failures in real-estate ventures in Asia (Williams 2016). The branding of new towns as smart cities is a general trend worldwide to sell often speculative development visions, as various incentives currently suggest in India (Praharaj & Han 2019).

In general, it can be summarised that contemporary new town master planning is possibly one of the most debated applications of urban design, which is still rather dominated by the image production of a new brand signalling growth and prosperity rather than the ideals of participation and integration requiring slower developments. Instead, new towns have often become an essential part of stimulating economic growth, especially in developing countries with an access to young populations and many low-income labourers, who can be engaged in construction industries. Most new town projects are facing major challenges in their long-term development, since they were initiated by a real estate-driven incentive and an often-vague idea of attracting new industries and service sectors in future. The integration of public employers, such as universities or ministries, can on the one side have a positive effect on economic incentives in emerging satellite or edge cities. On the other side, this strategy can foster further social segregation. In general, it can be observed that many recent peripheral new towns (e.g. New Cairo, New Mumbai) have been positioned in proximity to international airports. Their initiation has been following both growth incentives and the rationale of providing a safer and more comfortable living environment for higher-income groups to avoid the known effects of losing knowledge workers to international migration.

The vision of managing growth via new green, safe, and smart environments has however led to a rather complex role of master planning being a vehicle of communicating futuristic urban landscapes for yet unknown communities. Today, clients of new town projects can vary from governmental to major private sector stakeholders and often combinations of both in form of public–private partnerships. The main challenge for the commissioned urban designer is to identify the future residing communities and their ways of living to negotiate more directly communicated market requirements. This mainly concerns strategies for a more walkable environment to promote car independence as well as a more flexible and collaborative formation of future neighbourhoods via shared spaces. New town projects are often initiated in former unbuilt areas, such agricultural land, and in proximity to already existing infrastructure and smaller settlements, which are often integrated into the new spatial planning approach. The project type is thus rather related to the second type, introduced as greenfield settlements. The biggest difference between both types is the overall vision of establishing a new image of a town rather than an appendix in the periphery of an existing city.

Greenfield settlements are therefore usually managed by regional and municipal governance in collaboration with private sector stakeholders and are mainly focusing on urban peripheries. These settlements can often have very similar characteristics to new town projects, but their predominant focus on adjacent greenfield sites in urban peripheries is leading to a bigger dependence on existing urban centres. They are thus often referred to as new districts or dormitory settlements and again have a long history of planning experiences during various distinctive phases. One phenomenon of these settlements was experienced in the mass housing schemes of the post-war era, which followed modernist ideals of affordable living in large complexes surrounded by shared green spaces.

These settlements, such as the Banlieues of Paris or the Bijlmermeer development in Amsterdam (Helleman & Wassenberg 2004), led to the challenge of social disintegration due to many lower-income groups being moved to outer peripheries with limited access to local markets and infrastructure. The concentration of deprived communities has furthermore led to an intensification of social problems in a rather specific spatial configuration of mass housing typologies. While the modernist urban design was driven by the ideas of communal living and abstract open geometries, the missing understanding of how these large-scale typologies impact the general psychology of residents led to spaces associated with anonymity rather than community. In combination with the residents' economic vulnerability, the typologies exacerbated the overall experience of being a small component with limited choices instead of becoming an active part of a forming neighbourhood. Thus, planning and designing residential settlements need to be seen as a particular challenge, since historically residential settlements were often collectively built and formed, mainly defined by infrastructural supply and economic means. This can be still observed by sprawling urban agglomerations in developing countries, in which basic infrastructural planning is leading to ad hoc settlements being self-built via basic low-rise dwellings (Shao et al. 2020).

In Europe, the first period of large-scale modern planning during the dawn of the second Industrial Revolution led to strategic urban expansion plans on greenfield territory with clear typological decisions defining latter urbanism. On some occasions, these expansion plans even exceed the territories assigned to entire new town projects. One example of these historic master plans in form of urban expansions strategies was the so-called *Hobrecht* plan in Berlin, introduced in 1862, which defined a new grid, zoning policies, and a high-density perimeter block typology, known as *Mietskaserne*, to accommodate masses. This led to a distinctive spatial experience of these new districts on prior agricultural land formed by boulevards and the regularity of big urban blocks with shared inner courtyards (Rousset 2021).

Another example of a rapid expansion in the nineteenth century is Barcelona, where the so-called *Cerdà* plan defined one regular perimeter block with distinctive dimensions and being copied in all key directions (Aibar & Bijker 1997). While there have been first zoning regulations, the nineteenth-century expansion plans were following the principles of incorporating first public transit, horse carriages, and primarily pedestrians. Thus, density and land-use

Figure 6.4 The block structure defined by nineteenth-century planning in Berlin (Germany).

Source: Photograph taken by the authors.

integration were prime planning concerns to enable short walking distances, while the integration of urban design was centred on street sections of main boulevards and the integration of bigger open spaces, such as plazas and parks. The urban design objectives focused on establishing key places and orientation by allocating spaces for leisure and new monuments in form of architectural landmarks. This trend of city beautification in the nineteenth century can be linked to urban traditions of placemaking rooted in the European Renaissance (Figure 6.4) (Argan, 1969).

This small excursion to the nineteenth century is a small reminder that those top-down master plans for urban expansions on greenfield areas have a long history and have always been focusing on managing housing supply and growth dynamics. The introduction of the car as new mode of transportation led to a critical transformation of urban planning being centred on defining suburban grids permitting lower densities and a functional land-use division. This focus on streets as functionally needed infrastructure rather than as shared social spaces led to the missing integration of urban design aspects, which were more proactively pursued in former centuries. In recent decades, the urban sprawl of most agglomerations has led to strong limitations of initiating new greenfield projects. To support a reduction of daily commuting and to enable more social integration, urban design strategies have been rediscovered as important basis for master planning greenfield projects in recent decades.

One example for a twenty-first-century greenfield expansion can be found in the North-West of Frankfurt am Main, Germany. The Riedberg settlement was initiated in the 1990s and the construction started in the early 2000s on a previously agriculturally used area of 267 hectares (Kaufmann & Peterek 2018). The main aim was to accommodate around 15,000 new residents via 6,000 housing units, while a university campus was added, and a tram line extended to enable a transit-oriented approach in the urban periphery. The master plan was mainly created by urban planners and designers at the local city council and integrated an open space strategy in form of public parks and a shared commercial centre, marked by a public plaza. The project is a typical precedent for contemporary approaches of greenfield extensions in a European context. These developments are not centred anymore on suburban low-rise typologies and instead mix higher densities to integrate a bigger share of affordable housing and to encourage the use of public transit, whose effective provision requires a big number of daily commuters. One of the main challenges of designing an urban extension on greenfield sites is to establish a distinctive place via shared spaces and infrastructure and to avoid a dominance of suburban typologies, which were traditionally faster to supply due to the main risk of investment being taken by individual families rather than by major developers.

Greenfield projects as part of integrated development strategies are thus mainly observed in urban agglomerations with significant economic importance and a rising housing demand by middle- to high-income groups. In developing regions, such as Mumbai, greenfield urban extensions are often carried out by private sector incentives and in form of major residential compounds. These typologies often reduce the design aspects to landscaping matters of privatised spaces between residential high rises and behind walls. Similar trends have been observed in Chinese mega cities (Wang & Wiedmann 2024). The result has often been a patchwork of compounds in urban peripheries creating a fragmented landscape of privatised areas, which can sometimes individually reach sizes of more than twenty hectares, leading to a restricted walkability along main grids. While master planning for major residential developments is more common in urban peripheries on greenfield territory, there has been a major shift in most development strategies to favour the conversion of underused land, such as brownfield sites. This third project type can be located in both urban peripheries and in central areas (Figure 6.5).

Figure 6.5 The typical walking experience in new edge cities in Beijing (China).

Source: Photograph taken by the authors.

One of the main tasks of planning authorities is to identify opportunity sites for new developments within already existing urban boundaries to avoid the expansion on greenfield sites. Since continuous urban sprawl will always lead to an increasing need for new infrastructure and a reduced balance between built and unbuilt land. This balance is needed to avoid urban heat islands to manage urban growth towards more integrated dynamics following the vision of a city with short paths between living, working, and leisure. The rules for declaring opportunity sites via rezoning can be easily explained by local market dynamics. In many developed urban agglomerations, manufacturing industries, especially industries contributing to air pollution, are given incentives to leave sites and to enable a redevelopment. Other examples can be found in the case of major infrastructures, such as previous city airports, ports, or train tracks, which have been replaced via new developments. The same can be found in form of major logistic warehouses or military compounds. In some developing regions, even previous low-rise residential areas are often considered for redevelopment due to their lower densities and outdated infrastructure. A specific case is the focus on waterfront sites, which were previously occupied by a mix of port infrastructure and logistics, and which have been discovered as very valuable new city districts offering opportunities for a large volume of residential use.

Examples can be found in Boston (Heeg 2008), the Nordhavn in Copenhagen (Grauslund Kristensen 2024), or the HafenCity in Hamburg (Bruns-Berentelg 2010). All these redevelopments share that planning authorities have been aiming for integrating land uses, while the surrounding open spaces of waterways enabled the decision for rather high densities. The proximity to existing city centres furthermore enabled the integration of new cultural venues and landmarks as well as an improved pedestrian and public transit access. In future, it can be expected that large conversions of industrial waterfronts and ports will be launched in Chinese urban agglomerations due to an expected economic transformation. Thus, the general trend of redeveloping more accessible and thus underused urban sites to reduce new infrastructural developments will remain a key market for urban design services. The main risks are however the increasing land value expectations of existing landowners and the major costs of cleaning brownfield areas from previous pollution. In the case of Liverpool Waters, commencing plans

Figure 6.6 New developments at Liverpool's historic port and waterfront (UK).

Source: Photograph taken by the authors.

and strategies for redevelopment of the port area have led to the loss of the UNESCO World Heritage status of the waterfront in 2021 (Hole & Alsalloum 2024). Thus, urban designers can be often challenged in conversion projects to find a balance between the old images of a historic site and a new modern skyline (Figure 6.6).

This implies that there is a rather direct transition from master planning to convert central urban sites to project type 4: urban regeneration and conservation. While both types can often be interrelated, there needs to be a rather clear distinction between the redevelopment of larger sites via traditional master planning, and more holistic urban regeneration frameworks with an emphasis on public spaces, new development policies, and infrastructure. This framework planning aims to identify both conservation sites and opportunity sites, and one of the main aims is to upgrade key corridors to establish connected places creating new synergies for local developments and urban heritage. In many regards, regeneration plans can have a much larger scope and comprise the entire city, while the project duration and budget are usually rather limited and thus often carried out by local urban authorities with occasional

Figure 6.7 High-rise buildings in the City of London (UK).
Source: Photograph taken by the authors.

support by external consultants. Urban heritage does not only consist of specific buildings, plazas, or landscaped areas, and it also refers to entire historical townscapes, which should not be interrupted or significantly transformed by new developments. Some city authorities have for instance decided to restrict the building heights in historic centres, as in the example of Munich, Germany. And other city authorities, such as in London or Frankfurt am Main, have decided to initiate strategic framework plans for the allocation of new high-rise buildings in form of specific clusters. High-rise towers and the resulting globalised images are widely seen as the most challenging new additions to any historic townscape (Figure 6.7).

Apart from identifying and listing urban heritage, defining opportunity sites and their policy including building heights, and creating new meaningful links via upgraded public realm, regeneration framework plans can include a variety of infrastructural and transport-related strategies. As early as in the 1960s, the problematic effects of car traffic were recognised by planners in historical European city centres. The planner Colin Buchanan (1963) was one of the first pioneers to enable an improved conservation of a pedestrianised city centre by constructing ring roads and by restricting traffic in key locations. This established strategy of preserving a

URBAN DESIGN AND GOVERNANCE 199

Figure 6.8 The pedestrianised Buchanan Street in Glasgow (UK).
Source: Photograph taken by the authors.

shared public centre free of traffic has however hardly been applied in most cities in the developing world. Reasons can be found in the rapid modernisation, the climatic conditions, and the lack of capacities to enforce car restrictions. The rediscovery of streets in city centres as uncompromised pedestrian spaces can play however one of the most significant roles in urban regeneration, since places are experienced via human exploration enabling not only a deeper connection to historical surroundings but also encouraging more diverse market activities (Figure 6.8).

The biggest threat for a more integrated local economic development enabling smaller businesses to grow can be identified in form of both large malls and online shopping trends. This phenomenon can be observed worldwide, and it has a particularly problematic impact in cities with an already existing high level of social segregation. The spatially divided income groups have often lost a shared central marketplace and city centres have often become increasingly abandoned due to their deterioration, inaccessibility due to congestion, and their subsequent lack of economic competitiveness. Urban regeneration plans have therefore become increasingly challenged to introduce new visions and to attract higher-income groups to rediscover areas as potential habitats. While gentrification is often discussed from a Western experience and the negative effects of lower-income groups being pushed out of their previous neighbourhoods, the complete absence of higher-income groups in central urban areas has proven to be as negative (Lees & Phillips 2018).

Another important concern of regeneration plans is the integration of better tourism strategies by locating hotels in specific areas, restricting the transformation of housing stock into temporary tourist flats, and by understanding the main movement corridors to avoid congestion. In summary, it can be stated that this project type is to be seen as a continuous effort of urban planners and designers, often in various and connected phases. In many cities, plans need to be updated at least every five years and need to be synchronised with all existing data sets. The most important challenge of these plans is their actual implementation via new policies directing investments due to their nature of often having only long-term positive effects on economic growth, once areas have been rediscovered as shared opportunities. Thus, restrictions on certain ad hoc developments are often required despite the need for new investment and preferences of individual landowners. It is therefore often recommended to identify pilot projects and to initiate a bigger platform for communicating new shared visions to enable a participatory long-term urban renewal.

The last project type, which needs to be introduced and discussed, is the only urban design service, which is not bound on a specific site or area and only addresses the general local conditions of either a specific city, region, or an entire nation. The so-called urban design compendiums are elaborated recommendations and guidelines to inform future master planning on key design principles. Compendiums are seen as a concise summary of scientifically and precedent-informed best practice approaches to aid better urban design in specific climatic and cultural contexts. The economic context is usually less addressed since it is more dependent on specific sites. Examples can be found in the Qatar Urban Design Compendium (Ministry of Municipality 2025), which addresses the specific challenges of a desert climate as well as the very specific contexts of vernacular urban typologies and their cultural meaning to encourage a better response of architects and urban designers to the existing local design languages rather than importing approaches.

A pioneering urban design compendium was the UK Urban Design Compendium, published by English Partnerships in 2000 and withdrawn in 2021. Another edition was elaborated by the Homes and Communities Agency in 2013. The compendium follows the approach to introduce the principles of creating a functioning urban structure from a macro perspective, especially focusing on the various connections, and then moving into the principles of placemaking. While it has certain roots in the UK context, it can however be seen as a rather general document and mainly rooted in disseminating the basic urban design approach of understanding the existing structure first before establishing new links and activating specific places.

To sum up all five project types representing urban design services, we can state that each project type has a distinctive objective and a particular background in being initiated by either public or private sector, or in some cases both. In any project type, urban designers need to collaborate with the public sector, as the policymaker. As urban design is concerned about the configuration of public spaces and thus the key links in our cities defining our everyday urbanism, the public sector and its local governance need to be ideally as much involved as possible to introduce a shared vision and to set guidelines and policies. In recent decades, the first three project types including new towns, greenfield projects, and larger conversions of underused areas have been increasingly governed by private sector interests due to the increasing need for initial investment including infrastructure. In many cases, there have been so-called master developers with a certain amount of public control, while the entire development approach has been following the rationale of feasibility and economic return in shorter periods of time. This has enabled faster deliveries of large-scale mega projects, while it has also led to less measures for integration and long-term consolidation. Major compromises can often be found in social inclusion matters due to a missing mix of typologies, better qualities of public spaces, and higher levels of connectivity to diversify typologies, uses, and modes of transportation.

While site-specific master planning has become increasingly dominated by private sector initiatives and the general mechanisms of approval by local administrations, the last two project types are mainly located in the realms of governmental institutions and their planning authorities. Consulting services for these two project types are frequently outsourced, following the general procedure of public bidding by various urban design service providers. Both regeneration plans and design compendiums however require a close collaboration with urban designers engaged in local planning authorities to address all major concerns and to enable a future implementation in everyday urban governance. The typical urban designer is thus offering a wide range of services: from site analyses to master planning and from various urban studies to policy recommendations and design guidelines.

PROJECT MANAGEMENT

Any urban design project is starting with the basics of proactive communication, which includes a comprehensive exploration and acquisition of essential information. At first, an urban design project is identified by the need for an improved public space, which is any land that cannot be privately owned and thus mainly consists of public streets. These connections are however largely defined by surrounding buildings and their interaction. And vice versa the public space is the essential link enabling access and activating buildings, resulting markets, and forming communities. Public space deficits can be found in any existing environment requiring upgrading and on bigger pieces of land facing an entire redevelopment. The need for designing these shared public spaces is usually identified by stakeholders in the local urban authorities and their planning departments, who are surveying the status of the built environment and declaring new opportunity sites. This information gathering and dissemination are the foundations for initiating new urban design projects, which are then either carried out by local authorities and their employed urban designers or commissioned to externals via public tender. In the case of site-specific master planning services, the private sector or public–private partnerships can be the main clients of commissioned urban design services.

The best practice for investigating first design visions for larger project sites is the initiation of design competitions during the first project stage of communication and exploration. Despite the well-known fact that design competitions have a long history of positively enabling a broader communication and thus awareness of various opportunities for specific sites, they have been struggling in their implementation due to capacity problems worldwide. These capacity problems can be identified in form of small local authorities, regulatory challenges, and time as well as financial constraints. To set up a productive design competition, an experienced urban design service provider or local authorities and their inhouse urban designers need to carry out all essential site analyses and key negotiations with developers and investors regarding main planning aspects, such as land uses, infrastructure, and built densities. This project stage is a core requirement to set up the basic planning and design parameters to communicate the shared briefing document for competing design practices.

In general, it can be observed that design competitions are most successful in their outcomes if they are based on detailed briefs to test a variety of innovative but comparable approaches. To be more time efficient, it is often recommended to invite a specific range of experienced urban designers to compete, who have already delivered successful master plans within the same project category and who can further inform the project with gained approaches and insights. In many cases, favoured master plans have been revised by integrating aspects from other proposals. The public access and dissemination of design competitions is a further beneficial factor in transparently communicating the new vision for a site with major impact on existing urban surroundings and to balance public impressions of top-down and purely investment-driven incentives.

Establishing a design competition culture in the context of site-specific master planning is furthermore a core requirement to nurture the knowledge production in the design service industry and their continuous professional development enabling a bigger market share of smaller but talented service providers.

The communication phase of any urban design project can be seen as the most critical phase due to the big need for explorations to understand both the function and form of a specific design intervention. Since the main aim is always centred on establishing the best possible spatial experience in public spaces, all main stakeholders must be identified and ideally integrated in the process of decision-making. From a private sector perspective, new developments must be feasible and ideally profitable. The private sector itself consists of a variety of stakeholders with different and sometimes even opposing interests. While main stakeholders can be identified in form of landowners, investors, and their developers, a variety of other stakeholders, such as contractors and infrastructure suppliers, can have a decisive impact on projects due to their capacities and capabilities. Another factor can be landowners or businesses in the case of adjacent areas who might be affected by new developments. Similarly, the public sector cannot be reduced to planning authorities and policymakers, who act in public interests. The surrounding communities of developments must be carefully considered, since their everyday life might be transformed, and new communities might arrive with new spatial practices and demands on services.

Thus, it is often common practice to install forums for representatives of private and public sector to communicate projects and their impact. This participatory approach to urban planning and design can be enriching in developing the briefing documents and collaboratively evaluating competition results. The more transparent major developments are communicated, and the more synergies can often be found for a more effective and efficient transformation of entire urban areas to establish win–win situations. This however requires competent communication skills from urban designers, who do not only need to share visions in form of master plans, presented in scaled plans and models, but also in-depth site and precedent studies via diagrams to explain the specific spatial context and its opportunities as well as constraints.

Site studies are therefore the very first step in any urban design project and need to be managed methodologically by starting on investigating the macro context first before assessing the actual site and its surroundings.

The macro context of any site is the entire city, and to analyse the current development trajectory of a city, secondary data need to be gathered and explored first. This secondary data contain all main planning documents, census data, and basic economic and environmental aspects to understand the current position of a city and its core challenges in establishing better forms of urbanism. The next step would be a more rigorous examination of the district area, in which the site is located, and a first stakeholder analysis of the various interests in the new development including a more detailed community profile (demographic and economic census data). The last step is the detailed study of the site and its adjacent surroundings by exploring the functional aspects starting with the actual dimensions of the site and the various forms of access. After understanding the locational matters, the various existing land and building uses including infrastructure can be explored in detail before resulting environmental factors are assessed. The environmental impact study of a site is usually first concerned with the impact of traffic (e.g. noise and air pollution) before assessing potentially polluted land and water, flood concerns, as well as existing green spaces in surrounding areas, their role in reducing urban heat islands and contributing biodiversity.

After the assessment of functional aspects of a site and how the site is currently integrated in the overall environmental performance of the district, the form aspects can be investigated. Form refers to the language of a place and what it culturally represents. Thus, heritage needs to be identified as well as core narratives around the site, which might require a certain response and integration. This dimension distinguishes urban design site investigations from most planning approaches, since the everyday experiences and the resulting imageability of a site are taken as important as functional matters to inform future spatial configurations. Beyond heritage aspects, the general languages of both landscapes and architecture are collected and evaluated to enrich the understanding of the site as a lived place via a qualitative approach. This also includes the exploration of existing cultural spatial practices in the area with significant meanings for a formed place attachment to understand their importance and to enable a future integration. Examples can be found in certain traditional markets on weekends or seasonal festivals, which require certain spatial settings.

After understanding this big variety of aspects of any site, the urban designer needs to able to visually communicate the findings via legible diagrams, plans, and models. At this stage, a concise SWOT analysis can help to initiate a shared awareness of the present strengths and weaknesses of a site by also enabling a first discussion of future opportunities while being conscious about potential threats. One best practice for a participatory approach is to initiate a public forum with invited representatives of all main stakeholders to present the outcomes of this study and to gather initial views on potential future developments and thus better insights on the various perceptions and perspectives. After this event, a detailed briefing document can be developed with the main client, and ideally, a design competition can be launched to transitions to a distinctive second phase: the masterplan. In conclusion, an ideal urban design project should be structured in two parts. The elaboration of a comprehensive brief, which contains the actual site study phase resulting in design parameters, followed by the master planning phase, if possible, in form of a competition, and its detailed implementation including policies.

These stages are however often not distinctively applied, and it is rather common that projects are directly commissioned by master developers to get faster approval for development. It is thus often up to national or regional authorities to set up the procedure and to ensure a more rigorous and transparent planning and design process. One example is the enforcement of design competitions being a legal requirement in the case of sites beyond a specific size. It is however worth noting that the need for investments and the on-going pressure for economic competitiveness of cities has been main factors for many urban design matters being managed from a rather limited perspective. A more proactive and communicative approach can however unlock synergies and more investors to join. This again depends on a bigger awareness of urban design being a key factor in improving our quality of urban lives and subsequent economic consolidation processes less dependent on physical growth incentives.

FURTHER READINGS

Adhya, A. & Plowright, P.D. (2022). *Urban Design made by Humans: A Handbook of Design Ideas*, Routledge: London.

Devish, O., Huybrechts, L. & De Ridder, R. (2019). *Participatory Design Theory: Using Technology and Social Media to Foster Civic Engagement*, Routledge: London.

Holdo, M. (2024). *Participatory Spaces under Urban Capitalism: Contesting the Boundaries of Democratic Practices*, Routledge: London.

Sanchez, J. (2021). *Architecture for the Commons: Participatory Systems in the Age of Platforms*, Routledge: London.

SUMMARY

In this final discussion chapter, urban design basics are introduced by providing a discourse on urban design practice as being an important part of urban governance. The main purpose of urban design interventions is centred on forming our everyday urbanism, the way we live in and experience cities. Beyond the introduction of increasing AI-supported applications, urban design will remain rooted in human-to-human communications to find ways to improve the relationship between our public realm, our shared places and connections, and thus our lives. There has been an on-going and unstoppable transformation in our cities, which is not only driven by technological progress, but also the interconnected social, economic, and environmental realities, experienced by all of us. These experiences are shared and form the basic attachment or alienation between cities and us. The more we perceive cities as being machines, which need to be engineered for better functioning lives, the more we will mainly focus on technological interventions. This technological progress and continuous modernisation can however only be achieved in shorter periods by enabling more growth incentives. The result has been an on-going conflict between growth management challenges and the need for modern infrastructure to become part of an increasingly globalised network of cities.

While many technological applications have undoubtedly improved our everyday lives, the accelerated production and consumption have had an enormous impact on our spatial realities and our socio-spatial relationships. Urban governance worldwide has become an increasingly abstract decision-making enterprise due to simplified ideas of public interests and dominant private sector incentives. This reductionism of perceiving inhabitants as moving data, representing abstract consumers and producers, has led to systems of decision-making which are focused on managing basic needs as necessary side effect of growth incentives. This has lowered the opportunities for a more bottom-up and participatory form of governance rooted in demand-driven dynamics and thus, inhabitants, who are perceived as voices and ideas rather than reduced data sets. Consequently, one of the biggest questions of urban governance and managing our urban spaces via urban design interventions is a renewed question of ownership. If we all own our cities, we are all responsible for their well-being and a conscious self-management of our urban lives will naturally lead to an enhanced interest in urban habitats being less defined by external and rationalised decision-making.

7

URBAN DESIGN
Creating a Holistically Healthy City

This book attempts to cover all the main basics of urban design, as an applied science, which is deeply rooted in the field of urban studies while being connected to architecture, urban planning, and landscape design with all of their historic traditions. In addition to this interdisciplinary scope of urban design, the international perspective can be seen as another challenge of such an endeavour, since urban design approaches depend on specific local contexts, such as culture and climate. While these local contexts differ from place to place, any urban design intervention should share one main objective: to help places to become active places for a healthier society. In recent years, we have become increasingly aware that the modern urban crisis, first acknowledged and mentioned in the 1950s, has developed into many parallel existing and interconnected crises: from housing crises to social segregation; fragmentation and isolation trends; to urban heat islands; air pollution; and the deterioration of city centres due to economic transformation and decline. Cities are still some of our biggest achievements as civilisation, but the perception of them as hubs for individual success rather than as shared habitats has led to an on-going privatisation and fragmentation of urban spaces.

Today, all crises accumulate into one bigger crisis, our public health crisis. Each single aspect of the various conflicts in our cities has been contributing to a challenging condition of our public health. For decades, cities have been the suppliers of healthier living conditions due to better access to infrastructure and diverse

DOI: 10.4324/9781003251200-7

marketplaces. The new development trajectory however suggests that cities are increasingly struggling in providing healthy living conditions due to overpopulation, outdated infrastructure, and a problematic globalised and industrialised agricultural production. The latter is often seen as a separate global challenge, but as Jane Jacobs (1970) and others have pointed out, it is a responsibility of us as an urbanising world population to carefully consider how resources are produced and used. The dependence on extensive mass production has led to widespread monocultures and the overuse of soil, which has been facing a serious rate of nutrient depletion. Indeed, as with other civilisations before us, we are facing the challenge of collectively sustaining fragile ecosystems by overcoming our individualistic tendencies of hiding behind our own limitations rather than searching for possibilities of change.

Public health can be seen as a rather wide scientific discipline on the crossroads, between applied sciences and social sciences. Urban design is an applied science, but due to its roots in urban studies and urbanism, and the way we are living in cities, it is directly connected. Public health depends on a few fundamental requirements: the physical needs of human beings in form of clean water and air; as well as sufficient shelter; nutritious food; and the possibility for everyday walks and exposure to daylight. On top of these basic needs, we can distinguish various needs required for our mental health including sufficient spaces for social interaction, general economic and social stability for individual opportunities, as well as place attachment via aesthetic and familiar environments. Urban design can have an impact on all of these factors. Indeed, an urban reconfiguration can support a better integration of activities and open spaces to reduce commuting and enable new modes of transportation for more time, cleaner air and for shared public and green spaces. These green spaces could also be increasingly used to integrate urban farming opportunities. In addition, densities could be more balanced to enable new and innovative housing models (Figures 7.1 and 7.2).

As already established in the previous chapters, urban design measures are always attempting a form of placemaking to engage people and to thus enhance place attachment. Social interaction in shared and public spaces is a key foundation for more local markets and the ability of smaller entrepreneurial incentives to diversify

Figure 7.1 The public realm is increasingly important in promoting physical, mental, and spiritual well-being (Tainan, Taiwan).

Source: Photograph taken by the authors.

Figure 7.2 The ability to play and interact in public space helps to attract people of all ages to use the public realm and helps to build a sense of community (Granary Square, London, UK).

Source: Photograph taken by the authors.

local economies. This economic stability is a major factor in any social stability, which is again an important need for a healthy environment and the ability for young generations and disadvantaged social groups to interact safely. While it is still rather common to often rush or exclude more extensive urban design interventions worldwide, this book needs to conclude with this important discourse on public health to increase a shared awareness on the importance of this discipline. Placemaking is far more challenging than place branding, and urban design has for too long been part of a commercialisation trend of urban spaces rather than their reconnection to people and their needs.

A starting point of urban design initiatives for better public health will be the focus on our neighbourhoods, as the most important building blocks of our cities. Neighbourhoods can only function if neighbours interact with each other and explore shared interests, such as the safety of their neighbourhood streets. From safety matters to comfort and aesthetic concerns, neighbourhood streets can be reinvented as important links to other neighbourhoods and thus the entire city. Once neighbourhoods are not perceived anymore from the prioritised perspective of safety, other dimensions, such as aesthetic matters can play an increasing role encouraging a more daily use of public spaces. Small pubs, grocery stores, or coffee shops can benefit from this increasing social engagement and add a shared marketplace, which needs to be seen as an asset for creating an active neighbourhood. In addition to being local businesses, these are important social hubs where local information is exchanged, and social relationships are formed leading to informal networks of support. This leads to more social resilience and more local capabilities of raising young people to be responsible parts of society (Figure 7.3).

Especially, young and old generations as well as vulnerable social groups can benefit from more local incentives that reduce social isolation. Public health will increasingly be challenged by mental health problems, which are the result of various factors, such as social isolation and loneliness but also economic decline and unemployment. The already commencing economic transformation fuelled by an increasing digitalisation and internationalisation will lead to less opportunities for service sector occupations in traditional structures. One opportunity can, however, be found in

Figure 7.3 Creating opportunities for exercise and play in public spaces are an important part of promoting healthy and cohesive communities (Wenxin Park, Shenzhen, China).

Source: Photograph taken by the authors.

more connectivity and information access for smaller start-up companies, especially in more creative and producing industries. These start-ups require spaces that encourage social interaction and a bigger exchange of shared interests, skills, and knowledge. Any economic restructuring can be seen as both a threat for traditional pathways to livelihoods and a potential to become an active part of a new economy. This however requires spatial configurations that enable ways of social interaction and shorter paths between living and working. While digital connections can enable the fast exchange of information, many studies have shown the importance of human-to-human interaction in forming synergetic relationships.

These relationships are the roots for new ideas and innovations, as well as opportunities to increase efforts in sharing modes of transportation, services, and workplaces as well as more flexible housing solutions. The vision of 15-Minute Cities is introduced in this book, but like the idea of Smart Cities, these concepts can only be implemented in convincing and feasible dimensions, if the groundwork is forming and developing. This groundwork is to be seen in a network of people and their interactions. Importantly, for this groundwork to emerge, we need urban design to help create public spaces which are safe, engaging, and attractive. So, the key to all solutions is the intertwined relationship between urban design and public health to enable the recapturing of local streets as shared spaces rather than as abstract infrastructure for private matters. This, however, implies major urban design interventions and opportunities for more designers to initiate a discourse with communities to co-decide new ways for their urban living. Urban design, as part of inter- and multi-disciplinary partnerships, will therefore need to become increasingly participatory to shift its focus from place branding for growth incentives to placemaking for healthy cities. This participation requires critical communicators, who can assess local contexts by being globally aware of our societal challenges.

FURTHER READINGS

Through *Urban Design: The Basics*, we hope that we have inspired readers to want to study the discipline further. We have, therefore, guided the reader towards more in-depth reading to enable a deeper understanding of the many topics discussed in the book. Here we identify some broader reading, in terms of books, journals, and web resources, that will enable readers to progress their knowledge of urban design and hopefully inspire them to explore the subject in more detail.

BOOKS

Barnett, J. (2023) *Implementing Urban Design: Green, Civic, and Community Strategies*, Routledge: New York.

Black, P., Martin, M., Phillips, R. & Sonbli, T. (2024) *Applied Urban Design: A Contextually Responsive Approach*, Routledge: New York.

Carmona, M. (2021) *Public Places Urban Spaces: The Dimensions of Urban Design*, Third Edition, Routledge: New York.

Cowan, R. (2021) *Essential Urban Design: A Handbook for Architects, Designers and Planners*, RIBA Publishing: London.

Matan, A. & Newman, P. (2017) *People Cities: The Life and Legacy of Jan Gehl*, Island Press: Washington DC.

Sim, D. (2019) *Soft City: Building Density for Everyday Life*, Island Press: Washington.

JOURNALS

Journal of Urban Design: www.tandfonline.com/journals/cjud20
Urban Design International: www.palgrave.com/gp/journal/41289
Urban Design Journal: www.udg.org.uk/publications/journal
Urban Design and Planning: www.emeraldgrouppublishing.com/journal/jurdp

OTHER RESOURCES

Gehl People: www.gehlpeople.com
Project for Public Spaces: www.pps.org
Urban Design Group: www.udg.org.uk
Congress for New Urbanism: www.cnu.org

BIBLIOGRAPHY

Abbott, A. (1997) Of Time and Space: The Contemporary Relevance of the Chicago School, *Social Forces*, 75(4), 1149–1182. https://doi.org/10.2307/2580667

Adhya, A. & Plowright, P.D. (2022) *Urban Design Made By Humans: A Handbook of Ideas*, Routledge: New York.

Ahlava, A. & Edelman, H. (2009) *Urban Design Management: A Good Practice Guide*, Taylor & Francis: London.

Ahuti, S. (2015) Industrial Growth and Environmental Degradation, *International Education and Research Journal*, 1(5). Retrieved from https://ierj.in/journal/index.php/ierj/article/view/46

Aibar, E. & Bijker, W.E. (1997) Constructing a City: The Cerdà Plan for the Extension of Barcelona, *Science, Technology, & Human Values*, 22(1), 3–30, www.jstor.org/stable/689964.

Aldous, T. (1992) *Urban Villages: A Concept for Creating Mixed-Use Urban Developments on a Sustainable Scale*, Urban Villages Group: London.

Alexander, A. (2009) *Britain's New Towns: Garden Cities to Sustainable Communities*, Routledge: London.

Alexander, C. (1965) A City is Not a Tree, *Architectural Forum*, 122(1): 58–62 (Part I) April, and 122(2): 58–62 (Part II) May.

Alexander, C. (1979) *The Timeless Way of Building*, Oxford University Press: New York.

Alexander, C. (2002–2004) *The Nature of Order: An Essay on the Art of Building and the Nature of the Universe*, Series of Four Books, Center for Environmental Structure: Berkeley.

Alexander, C., Ishikawa, S., Silverstein, M., Jacobson, M., Fiksdahl-King, I. & Angel, S. (1977) *A Pattern Language: Towns, Buildings, Construction*, Oxford University Press: New York.

Alexander, C., Neis, H. & Alexander, M.M. (2013) *The Battle for the Life and Beauty of the Earth: A Struggle Between Two World Systems*, Oxford University Press: New York.

Alexander, C., Neis, H., Anninou, A. & King, I. (1987) *A New Theory of Urban Design*, Oxford University Press: New York.

Alexander, C., Silverstein, M., Angel, S., Ishikawa, S. & Abrams, D. (1975) *The Oregon Experiment*, Oxford University Press: New York.

Alvarez, L. (2023) *A Beginner's Guide to Urban Design and Development: The ABC of Quality, Sustainable Design*, Routledge: New York.

American Planning Association, Steiner, F. & Butler, K. (2007) *Planning and Urban Design Standards*, John Wiley: New York.

Appleyard, D., Gerson, S. & Lintell, M. (1981) *Livable Streets*, University of California Press: Berkeley.

Arfa, F.H., Zijlstra, H., Lubelli, B. & Quist, W. (2022) Adaptive Reuse of Heritage Buildings: From a Literature Review to a Model of Practice, *The Historic Environment: Policy & Practice*, 13(2), 148–170. https://doi.org/10.1080/17567505.2022.2058551

Argan, G.C. (1969) *The Renaissance City, Planning and Cities*, George Braziller: New York.

ARUP (2016) *Cities Alive: Towards a Walking World*, 27 April, ARUP: London.

ARUP (2020) *Transform and Reuse: Low-Carbon Futures for Existing Buildings*, ARUP: London.

ARUP (no date) *Delivering Sustainably Walkable Neighbourhoods: An Evidence-led Approach*, www.arup.com/insights/delivering-sustainably-walkable-neighbourhoods (accessed 20 September 2024).

ARUP (no date) *Walkable Cities are Better Cities*, www.arup.com/perspectives/walkable-cities-are-better-cities (accessed 23 June 2023).

Bacon, E.N. (1975) *Design of the Cities*, Revised Edition, Thames & Hudson: London.

Baker-Brown, D. (2024) *The Re-Use Atlas: A Designer's Guide Towards a Circular Economy*, RIBA Publishing: London.

Banerjee, T. & Loukaitou-Sideris, A. (eds.) (2019) *The New Companion to Urban Design*, Routledge: Oxford.

Barnett, J. (2023) *Implementing Urban Design: Green, Civic, and Community Strategies*, Routledge: New York.

Barton, H., Grant, M. & Guise, R. (2021) *Shaping Neighbourhoods for Local Health and Global Sustainability*, Routledge: London.

Barz-Malfatti, H. & Signer, S. (2020) *New Public Spaces – European Urban Squares in the 21st Century*, M BOOKS: Weimar.

Batty, M. (2016) Big Data and the City. *Built Environment*, 42(3), 321–337, www.jstor.org/stable/44132282

Bentley, I., De, S., McGlynn, S. & Rampuria, P. (2024) *EcoResponsive Environments: A Framework for Settlement Design*, Routledge: London.

Bentley, I., McGlynn, S., Smith, G., Alcock, A. & Murrain, P. (1985) *Responsive Environments: A Manual for Designers*, Routledge: Oxford.

Bishop, P. & Williams, L. (2012) *The Temporary City*, Routledge: London.
Black, P. & Sonbli, T.E. (2019) *The Urban Design Process*, Lund Humphries: London.
Black, P., Martin, M., Phillips, R. & Sonbli, T. (2024) *Applied Urban Design: A Contextually Responsive Approach*, Routledge: New York.
Branch, M. (1985) *Comprehensive City Planning: Introduction & Explanation*, Routledge: London.
Bratishenko, L. (2016) Jane Jacobs's Tunnel Vision: Why Our Cities Need Less of Jane Jacobs, *Literary Review of Canada*, October. https://reviewcanada.ca/magazine/2016/10/jane-jacobss-tunnel-vision
Broadbent, G. (1995) *Emerging Concepts in Urban Design*, Taylor & Francis: New York.
Brown, R., Hanna, K. & Holdsworth, R. (eds) (2017) *Making Good – Shaping Places for People*, Centre for London: London, www.yumpu.com/en/document/read/57113685/making-good-shaping-places-for-people (accessed 27 July 2023).
Brundtland, G. (1987) *Report of the World Commission on Environment and Development: Our Common Future*, United Nations General Assembly Document A/42/427.
Bruns-Berentelg, J. (2010) *Hafen City Hamburg: Places of Urban Encounter Between Metropolis and Neighborhood*, Berlin: Springer Verlag.
Bruntlett, M. & Bruntlett, C. (2018) *Building the Cycling City: The Dutch Blueprint for Urban Vitality*, Island Press: Washington DC.
Bruntlett, M. & Bruntlett, C. (2021) *Curbing Traffic: The Human Case for Fewer Cars in Our Lives*, Island Press: Washington DC.
Buchanan, C. (1963) *Traffic in Towns: A Study of the Long-Term Problems of Traffic in Urban Areas*, HMSO: London.
Bullivant, L. (2012) *Masterplanning Futures*, Routledge: London.
Burgess, E.W. (2008) The Growth of the City: An Introduction to a Research Project, In. Marzluft, J.M., *et al. Urban Ecology*. Springer: Boston, MA.
Burke, S. (2016) *Placemaking and the Human Scale City, Project for Public Spaces*, www.pps.org/article/placemaking-and-the-human-scale-city
Bürklin, T. & Peterek, M. (2007) *Basic Urban Building Blocks*, Birkhäuser: Basel.
Burns, M. (2020) *New Life in Public Squares*, RIBA Publishing: London.
Burton, E., Jenks, M. & Williams, K. (eds) (1996) *The Compact City: A Sustainable Urban Form?*, London: Routledge.
Burton, E., Jenks, M. & Williams, K. (eds) (2000) *Achieving Sustainable Urban Form*, Routledge: London.
Burton, E. & Mitchell, L. (2016) *Inclusive Urban Design: Streets for Life*, Routledge: London.
CABE (2005) *Better Neighbourhoods: Making Higher Densities Work*, Commission for Architecture & the Built Environment: London.

Campbell, C.J. (2018) Space, Place and Scale: Human Geography and Spatial History in Past and Present, *Past & Present*, 239(1): 23–45, https://doi.org/10.1093/pastj/gtw006

Carmona, M. (2019) Place Value: Place Quality and Its Impact on Health, Social, *Economic, and Environmental Outcomes, Journal of Urban Design*, 24(1): 1–48. https://doi.org/10.1080/13574809.2018.1472523

Carmona, M. (2021) *Public Places Urban Spaces: The Dimensions of Urban Design*, Third Edition, Routledge: New York.

Carmona, M., Bento, J. & Gabrieli, T. (2023) *Urban Design Governance: Soft Powers and the European Experience*, UCL Press: London.

Carmona, M. & Wunderlich, F.M. (2012) *Capital Spaces: The Multiple Complex Public Spaces of a Global City*, Routledge: London.

Case, B. & Van der Weele, T.J. (2024) Integrating the Humanities and the Social Sciences: Six Approaches and Case Studies, *Humanities and Social Sciences Communications*, 11. https://doi.org/10.1057/s41599-024-02684-4

Castells, M. (1978) Urban Crisis, Political Process and Urban Theory. In *City, Class and Power*, Palgrave: London.

Castells, M. (1983) *The City and the Grassroots: A Cross-cultural Theory of Urban Social Movements*, University of California Press: Berkeley.

Chamorro-Premuzic, T. (2023) *I, Human: AI, Automation, and the Quest to Reclaim What Makes Us Unique*, Harvard Business Review Press: Boston.

Chase, J., Crawford, M. & John, K. (2008) *Everyday Urbanism: Expanded*, The Monacelli Press: New York.

Chen, F. & Thwaites, K. (2016) *Chinese Urban Design: The Typomorphological Approach*, Routledge: London.

Chen, T., Ramon Gil-Garcia, J. & Gasco-Hernandez, M. (2022) Understanding Social Sustainability for Smart Cities: The Importance of Inclusion, Equity, and Citizen Participation as both Inputs and Long-term Outcomes, *Journal of Smart Cities and Society*, 1(2):135–148. https://doi.org/10.3233/SCS-210123

Chicago, M.P.C. (2008) *A Guide to Neighbourhood Placemaking in Chicago*, Project for Public Spaces and Metropolitan Planning Council: Chicago.

Childs, M.C. (2006) *Squares: A Public Place Design Guide for Urbanists*, University of New Mexico Press: Albuquerque.

CNU (2024) *Congress for New Urbanism: The Movement*, www.cnu.org/who-we-are

CNU & Talen, E. (ed) (2013) *Charter of the New Urbanism*, 2nd edition, McGraw Hill New York.

Coates, A. (2023) How Do Philosophical Positions Influence the Social Science Research Process? A Classification and Metaphor Analysis of Researchers' Descriptions, *Social Epistemology*, 38(5), 632–650. https://doi.org/10.1080/02691728.2023.2283447

Cohen, S. & Taylor, L. (1993) *Escape Attempts: The Theory and Practice of Resistance in Everyday Life*, Routledge: London.

Collins, G.R. & Craseman-Collins, C. (2006) *Camillo Sitte: The Birth of Modern City Planning*, Dover Publications: New York.

Cook, I.R. (2009) Private Sector Involvement in Urban Governance: The Case of Business Improvement Districts and Town Centre Management Partnerships in England, *Geoforum*, 40(5), 930–940, https://doi.org/10.1016/j.geoforum.2009.07.003

Cookson Smith, P. (2023) *Writings on the Asian City*, ORO Editions: Hong Kong.

Corbett, N. (2004) *Transforming Cities: Revival in the Square*, RIBA Enterprises: London.

Coupland, A. (ed) (1997) *Reclaiming the City: Mixed Use Development*, Spon Press: London.

Cowan, R. (2021) *Essential Urban Design: A Handbook for Architects, Designers and Planners*, RIBA Publishing: London.

CPRE (2019) *Double the density, halve the land needed*, Report by Campaign to Protect Rural England, www.cprelondon.org.uk/wp-content/uploads/sites/10/2020/02/DoubleTheDensityHalveTheLandNeeded_1.pdf

Cullen, G. (1961) *The Concise Townscape*, Architectural Press: Oxford.

Cuthbert, A.R. (2003) *Designing Cities: Critical Readings in Urban Design*, Blackwell Publishing: Oxford.

Cuthbert, A.R. (2006) *The Form of Cities: Political Economy and Urban Design*, Blackwell Publishing: Oxford.

Cuthbert, A.R. (2011) *Understanding Cities: Method in Urban Design*, Routledge: London.

D'Acci, L. (2013) Simulating Future Societies in Isobenefit Cities: Social Isobenefit Scenarios, *Futures*, November, 53, 3–18, https://doi.org/10.1016/j.futures.2013.09.004

da Cruz, N.F., Rode, P., & McQuarrie, M. (2018) New Urban Governance: A Review of Current Themes and Future Priorities. *Journal of Urban Affairs*, 41(1), 1–19. https://doi.org/10.1080/07352166.2018.1499416

Dalsgaard, A. (2012) *The Human Scale: Bringing Cities to Life*, Final Cut for Real: Copenhagen.

Dantzig, G.B. & Saaty, T.L. (1973) *Compact City: A Plan for a Liveable Urban Environment*, W.H. Freeman & Co.: San Francisco.

Das, D.K. (2022) Appraisal of the Linkage among Urban Infrastructure and Human Resources and the Growth of Information Technology Industry in Indian Cities. *Cogent Engineering*, 9(1). https://doi.org/10.1080/23311916.2022.2034263

De Oliveira, N.G. (2020) *Mega-Events, City and Power*, Routledge: London.

DETR (2000) *By Design: Urban Design in the Planning System: Towards Better Practice*, Department of the Environment, Transport and the Regions, CABE: London.

DoE (1994) *Quality in Town and Country,* Department of the Environment, HMSO: London.

Dovey, K. (2016) *Urban Design Thinking: A Conceptual Toolkit.* London: Bloomsbury Academic, pp. 9–16. http://dx.doi.org/10.5040/9781474228503.ch-001

Duany, A. & Plater-Zyberk (1991) *Towns and Town-Making Principles,* Rizzoli: New York.

Duany, A., Plater-Zyberk, E. & Speck, J. (2000) *Suburban Nation: The Rise of Sprawl and the Decline of the American;* North Point Press: New York.

Edwards, M (2001) City Design: What Went Wrong at Milton Keynes? *Journal of Urban Design,* 6(1), 73–82.

Eisenman, P. & Iturbe, E. (2020) *Lateness,* Princeton University Press: Princeton.

Ellin, N. (2000) *Postmodern Urbanism,* Princeton Architectural Press: New York.

EPOA (2018) *The Essex Design Guide,* Essex Planning Officers Association, www.essexdesignguide.co.uk

Essex County Council (2021a) *The Essex Design Guide,* Essex County Council: Chelmsford.

Essex County Council (2021b) *The Essex Design Guide: Successful Criteria for Public Open Spaces,* Essex County Council: Chelmsford, www.essexdesignguide.co.uk/design-details/landscape-and-greenspaces/successful-criteria-for-public-open-spaces

Ewert, A., Milton, D. & Overholt, J. (2019) *Natural Environments and Human Health,* CABI Publishing: Wallingford.

Farr, D. (2018) *Sustainable Nation: Urban Design Patterns for the Future,* Wiley: New York.

Farrell, T. (2013) *The City as a Tangled Bank: Urban Design vs Urban Evolution,* AD Primers, John Wiley: Chichester.

Farrelly, L. (2011) *Drawing for Urban Design,* Laurence King Publishing: London.

Flanders, D. & Miller, E. (2020) *Creating Great Places: Evidence-based Urban Design for Health and Wellbeing,* Routledge: New York.

Florida, R.L. (2002) *The Rise of the Creative Class: And How It's Transforming Work, Leisure, Community and Everyday Life,* Basic Books: New York.

Forsyth, A. (2015) What is a Walkable Place? The Walkability Debate in Urban Design, *Urban Design International,* 20: 274–292. https://doi.org/10.1057/udi.2015.22

Foucault, M. (1972) *The Archaeology of Knowledge and the Discourse of Language,* Pantheon Books: New York.

Fowler, W.R. (2022) *A Historical Archaeology of Early Spanish Colonial Urbanism in Central America,* University Press of Florida: Gainesville.

Frederick, M. (2018) *101 Things I Learned in Urban Design School,* Three Rivers Press: New York.

Frederick, M. & Mehta, V. (2018) *101 Things I Learned in Urban Design School,* Three Rivers Press: New York.

Friedmann, J. (1986) The World City Hypothesis, *Development and Change,* 17, 69–83. https://doi.org/10.1111/j.1467-7660.1986.tb00231.x

Fujita, M., Krugman, P. & Mori, T. (1999) On the Evolution of Hierarchical Urban Systems, *European Economic Review,* 43(2): 209–251, https://doi.org/10.1016/S0014-2921(98)00066-X

Gardner, W. (2024) *How to Start Strengthening Your Town with Incremental Development, Strong Towns Journal,* May 6, www.strongtowns.org/journal/2024/5/6/how-to-start-strengthening-your-town-with-incremental-development (accessed 5th December 2024).

Gassner, G. (2019) *Ruined Skylines: Aesthetics, Politics and London's Towering Cityscape,* Routledge: London.

Geddes, P. (1904) *City Development: A Study of Parks, Gardens, and Culture-Institutes,* Geddes & Co.: Edinburgh.

Geddes, P. (1915) *Cities in Evolution: An Introduction to the Town Planning Movement and to the Study of Civics,* Williams & Norgate: London.

Geddes, P. (2019) *Cities in Evolution: An Introduction to the Town Planning Movement and to the Study of Civics,* Forgotten Books: London.

Gehl Institute (2017) *Inclusive Healthy Places: A Guide to Inclusion and Health in Public Space – Learning Globally to Transform Locally,* Gehl Institute with the Robert Wood Johnson Foundation, https://ihp.gehlpeople.com/wp-content/uploads/2022/08/Inclusive-Healthy-Places_Gehl-Institute.pdf (accessed 29 November 2024).

Gehl, J. (1996) *Public Spaces – Public Life,* Island Press: Washington DC.

Gehl, J. (2010) *Cities for People,* Island Press: Washington DC.

Gehl, J. (2011) *Life Between Buildings: Using Public Space,* 6th edition, Island Press: Washington DC.

Gehl, J. & Gemzøe, L. (2008) *New City Spaces,* Danish Architectural Press: Copenhagen.

Gehl, J. & Svarre, B. (2013) *How to Study Public Life: Methods in Urban Design,* Island Press: Washington DC.

Gehl People (2024) www.gehlpeople.com

George, H. (1880) *Progress and Poverty,* Appleton: New York, Retrieved from the Library of Congress, www.loc.gov/item/05022674

Gharipour, M. (ed.) (2016) *Contemporary Urban Landscapes of the Middle East,* Routledge: London.

Gill, T. (2021) *Urban Playground: How Child-Friendly Planning and Design Can Save Cities,* RIBA Publishing: London.

Glaesser, E. (2011) *Triumph of the City: How Our Best Invention Makes Us Richer, Smarter, Greener, Healthier, and Happier*, Penguin Press: New York.

Global Designing Cities Initiative (2016) *Global Street Design Guide*, Island Press: Washington. https://globaldesigningcities.org/publication/global-street-design-guide

Gold, J. (1997) *The Experience of Modernism: Modern Architects and the Future City*, Routledge: London.

Gosling, D. (1996) *Gordon Cullen: Visions of Urban Design*, Academy Editions: London.

Graham, G. (2006) *Philosophy of the Arts: An Introduction to Aesthetics*, Routledge: London.

Grauslund Kristensen, N. (2024) Consequential Urban Development of Sustainable Strategies. *Urban Geography*, 46(1): 231–248. https://doi.org/10.1080/02723638.2024.2359319

Greater London Authority (2016) *London Plan Density Research: Lessons from Higher Density*, GLA: London, www.london.gov.uk/sites/default/files/project_2_3_lessons_from_higher_density_development.pdf (accessed 25 August 2023).

Haas, T. (ed.) (2008) *New Urbanism and Beyond: Designing Cities for the Future*, Rizzoli: New York.

Hall, P. (1988) *Cities of Tomorrow: An Intellectual History of Urban Planning and Design in the Twentieth Century*, Blackwell: Oxford.

Hall, T. (1997) *Planning Europe's Capital Cities: Aspects of Nineteenth Century Urban Developments*, E & FN Spon: London.

Haneman, J.T. (1984) *Historical Architectural Plans, Details and Elements*, Dover: New York.

Hardy, T. (2012) *From Garden Cities to New Towns: Campaigning for Town and Country Planning 1899-1946*, Routledge: London.

Harvey, D. (1975) *Social Justice and the City*, Edward Arnold: London.

Harvey, D. (1981) The Urban Process Under Capitalism: A Framework for Analysis. In: Dear, M. & Scott, A. (eds), *Urbanization and Urban Planning in Capitalist Society*, Routledge: London.

Havinga, L., Colenbrander, B. & Schellen, H. (2020) Heritage Attributes of Post-War Housing in Amsterdam, *Frontiers of Architectural Research*, 9(1): 1–19. https://doi.org/10.1016/j.foar.2019.04.002

Hayward, R., McGlynn, S. & Reeve, A. (2010) *Better Towns and Cities: A Manual of Town Centre Management*, Architectural Press: Oxford.

Heeg, S. (2008) Property-Led Development Als Neuer Ansatz In Der Stadtentwicklung? Das Beispiel der South Boston Waterfront in Boston. *Erdkunde*, 62: 41–57.

Helleman, G. & Wassenberg, F. (2004) The Renewal of What Was Tomorrow's Idealistic City: Amsterdam's Bijlmermeer High-Rise, *Cities*, 21(1), 3–17, https://doi.org/10.1016/j.cities.2003.10.011

Henderson, J. (2020) EVs Are Not the Answer: A Mobility Justice Critique of Electric Vehicle Transitions, *Annals of the American Association of Geographers*, 110(6): 1993–2010. https://doi.org/10.1080/24694452.2020.1744422

Hickel, J. (2017) *The Divide: A Brief Guide to Global Inequality and its Solutions*, William Heinemann: London.

Higashide, S. (2019) *Better Buses Better Cities: How to Plan, Run, and Win the Fight for Effective Transport*, Island Press: Washington DC.

Hillier, B. (1996) *Space is the Machine: A Configurational Theory of Architecture*, Cambridge University Press: Cambridge.

Ho K.L. & Lee C. (2012) The Quality of Design Participation: Intersubjectivity in Design Practice, *International Journal of Design*, 6(1), www.ijdesign.org/index.php/IJDesign/article/view/749/404

Hochstenbach, C. & Musterd, S. (2017) Gentrification and the Suburbanization of Poverty: Changing Urban Geographies through Boom and Bust Periods, *Urban Geography*, 39(1), 26–53. https://doi.org/10.1080/02723638.2016.1276718

Hole, J. & Alsalloum, A. (2024) Evolution of Heritage and Development in Liverpool's Waterfront over 40 Years. *Discover Cities*, 1, 11, https://doi.org/10.1007/s44327-024-00012-8

Holmes, R. (1997) Genre Analysis, and the Social Sciences: An Investigation of the Structure of Research Article Discussion Sections in Three Disciplines, *English for Specific Purposes*, 16(4): 321–337, https://doi.org/10.1016/S0889-4906(96)00038-5

Hoshino, K. (1987) Semiotic Marketing and Product Conceptualization, In Umiker-Sebeok, J. (ed.) *Marketing and Semiotics: New Directions in the Study of Sign for Sale*, pp. 41–55, Mouton de Gruyter: Berlin.

Howard, E. (1898) *To-Morrow: A Peaceful Path to Real Reform*, Swan Sonnenschein: London. Republished as *Garden Cities of To-morrow* in 1902, Faber and Faber: London.

Hurtt, S., Latini, A.P. & Tiice, J. (eds.) (2025) *The Urban Design Legacy of Colin Rowe*, ORO Editions: Novato, CA.

Iovene, M., Boys Smith, N. & Seresinhe, C.I. (2019) *Of Street and Squares: Which Public Spaces Do People Want to Be In and Why?*, Cadogan: London.

Jacobs, A.B. (1993) *Great Streets*, MIT Press: Cambridge, Mass.

Jacobs, J. (1961) *The Death and Life of Great American Cities*, Random House: New York.

Jacobs, J. (1970) *The Economy of Cities*, Vintage Books: New York.

Jacobs, J. (1993) *The Death and Life of Great American Cities*, Vintage Books: New York.

Jacobs, J. (2011) *The Uses of Sidewalk: Safety*, In R.T. LeGates and F. Stout (eds.) *The City Reader*. 5th ed., pp. 105–109, Taylor & Francis: Abingdon.

Jacques, E., Neuenfeldt Júnior, A., De Paris, S., Matheus Francescatto, & Siluk, J. (2024) Smart Cities and Innovative Urban Management: Perspectives of Integrated Technological Solutions in Urban Environments, *Heliyon*, 10(6), https://doi.org/10.1016/j.heliyon.2024.e27850

Jones, P., Roberts, M. & Morris, L. (2007) *Rediscovering Mixed-Use Streets: The Contribution of Local High Streets to Sustainable Communities*, Policy Press: Bristol.

Kanigel, R. (2017) *Eyes on the Street: The Life of Jane Jacobs*, Vintage Books: New York.

Katz, P. (1994) *The New Urbanism: Toward an Architecture of Community*, McGraw Hill: New York.

Kaufmann, C. & Peterek, M. (2018) *Frankfurter Riedberg: Stadtentwicklung fuer das 21. Jahrhunder*, Jovis Verlag: Berlin.

Kiang, H.C., Liang, L.B. & Limin, H. (eds.) (2010) *On Asian Streets and Public Space*, NUS Press: Singapore.

Koch, R. & Latham, A. (eds) (2017) *Key Thinkers on Cities*, Sage Publications: Los Angeles.

Kohn, M. (2004) *Brave New Neighborhoods: The Privatization of Public Space*, New York: Routledge.

Kostof, S. (1991) *The City Shaped: Urban Patterns and Meanings Through History*, London: Thames & Hudson.

Kostof, S. (1992) *The City Assembled: The Elements of Urban Form Through History*, London: Thames & Hudson.

Krier, R. (2006) *Town Spaces: Contemporary Interpretations in Traditional Urbanism*, Birkhäuser: Basel.

Kuper, A. (1996) *The Social Science Encyclopedia*, Routledge: London.

LaFarge, A. (ed.) (2000) *The Essential William H. Whyte*, Fordham University Press: New York.

Lang, J. (1994) *Urban Design: The American Experience*, John Wiley: New York.

Lang, J. (2021) *The Routledge Companion to Twentieth and Early Twenty-First Century Urban Design: A History of Shifting Manifestoes, Paradigms, Generic Solutions, and Specific Designs*, Routledge: New York.

Lang, J. & Marshall, N. (2016) *Urban Squares as Places, Links and Displays: Success and Failures*, Routledge: New York.

Lange, S., Pohl, J. & Santarius, T. (2020) Digitalization and Energy Consumption: Does ICT Reduce Energy Demand?, *Ecological Economics*, 176, https://doi.org/10.1016/j.ecolecon.2020.106760

Larco, N. & Knudson, K. (2024) *The Sustainable Urban Design Handbook*, Routledge: New York.

Larice, M. & Macdonald, E. (eds) (2012) *The Urban Design Reader*, Routledge: London.

Larkham, P. & Conzen, M. (eds) (2014) *Shapers of Urban Form: Explorations in Morphological Analysis*, Routledge: New York.

Larson, K. (2012) *Brilliant Designs to Fit More People in Every City*, TedxBoston, June 2012, www.ted.com/talks/kent_larson_brilliant_designs_to_fit_more_people_in_every_city?language=en (accessed 21 August 2023).

Laurence, P. (2016) *Becoming Jane Jacobs*, University of Pennsylvania Press: Philadelphia.

Le Corbusier (1933) *La Ville Radieuse: Eléments d'une doctrine d'urbanisme pour l'équipement de la civilisation machinist*, London: Faber and Faber. Republished as *The Radiant City: Elements of a Doctrine of Urbanism to be Used as the Basis of Our Machine-Age Civilization (1967)*, Orion Press: New York.

Lees, L. & Phillips, M. (2018) *Handbook of Gentrification Studies*, Cheltenham: Edward Elgar Publishing Ltd.

Lefebvre, H. (1991) The Production of Space. Blackwell.

Lefebvre, H. (1996) The Right to the City, In Kofman, E. & Lebas, E. (eds.) *Writings on Cities*, Wiley-Blackwell: Cambridge, Mass.

Lefebvre, H. (2002) Comments on a New State Form, *Antipode*, 33(5): 783–808.

Lefebvre, H. (2003*)* *The Urban Revolution*, University of Minnesota Press: Minneapolis.

Lefebvre, H. (2009) *State, Space, World: Selected Essays*, University of Minnesota Press: Minneapolis.

Lefebvre, H. (2017) *Rhythmanalysis: Space, Time, and Everyday Life*, Bloomsbury: London.

Leffel, B. & Acuto, M. (2018) Economic Power Foundations of Cities in Global Governance. *Global Society*, 32(3): 281–301, https://doi.org/10.1080/13600 826.2018.1433130

Lerup, L. (1977) *Building the Unfinished: Architecture and Human Action*, Sage Publications: Beverly Hills.

Llewelyn-Davies (2000) *Urban Design Compendium, with English Partnerships,* The Housing Association: London.

Lloyd Wright, F. (1932) *The Disappearing City*, Farquhar Payson: New York.

Lock, K. & Ellis, H. (2020) *New Towns: The Rise, Fall and Rebirth*, RIBA Publishing: London.

Loew, S. (ed.) (2012) *Urban Design in Practice: An International Review*, RIBA Publishing: London.

Lofland, L. (1998) *The Public Realm: Exploring the City's Quintessential Social Territory*, Routledge: London.

Luscher, D. (n.d.) *The 15-Minute City Project,* www.15minutecity.com (accessed 21 August 2023).

Lynch, K. (1960) *The Image of the City*, MIT Press: Cambridge, Mass.

Lynch, K. (1984) *A Theory of Good City Form*, MIT Press: Cambridge, Mass.

Lynch, K. & Carr, S. (1996) Open Space: Freedom and Control. In T. Banerjee & M. Southworth (ed.) *City Sense and City Design: Writings and Projects of Kevin Lynch*, pp. 413 – 417, MIT Press: Cambridge.

Madanipour, A. (1996) *Design of Urban Space: An Inquiry into a Socio-Spatial Process*, John Wiley: New York.

Madanipour, A. (2017) *Cities in Time: Temporary Urbanism and the Future of the City*, Bloomsbury: London.

Madden, K. & Project for Public Places (2021) *How to Turn a Place Around: A Placemaking Handbook*, PPS: New York.

Malone, L. (2019) *Desire Lines: A Guide to Community Participation in Designing Places,* RIBA Publishing: London.

Marron, C. (ed) (2016) *City Squares: Eighteen Writers on the Spirit and Significance of Squares Around the World*, Harper Collins: New York.

Matan, A. & Newman, P. (2017) *People Cities: The Life and Legacy of Jan Gehl*, Island Press: Washington DC.

May, T. & Perry, B. (2018) *Cities and the Knowledge Economy: Promise, Politics and Possibilities*, Routledge: London.

Mead, G.H. (1934) *Mind, Self, and Society from the Standpoint of a Social Behaviorist*, University of Chicago Press: Chicago.

Medrano, L., Recamán, L. & Avermaete, T. (eds.) (2021) *The New Urban Condition: Criticism and Theory from Architecture and Urbanism*, Routledge: New York.

Meeda, B. (2018) *Graphics for Urban Design*, ICE Publishing: London.

Mehaffy, M.W. (2008) Generative Methods in Urban Design: A Progress Assessment, *Journal of Urbanism*, 1(1): 57–75, https://doi.org/10.1080/17549170801903678

Ministry of Housing, Communities & Local Government (2021) *National Design Guide: Planning Practice for Beautiful, Enduring and Successful Places*, MHCLG: London. https://assets.publishing.service.gov.uk/government/uploads/system/uploads/attachment_data/file/962113/National_design_guide.pdf (accessed 21 August 2023).

Ministry of Municipality, Qatar (2025) *Qatar Masterplan,* www.mme.gov.qa/QatarMasterPlan/English/strategicplans.aspx?panel=CompendiumProjects (accessed 16 January 2025).

Moilanen, T. & Rainisto, S. (2009) *How to Brand Nations, Cities and Destinations.* Palgrave Macmillan: London.

Molotch, H. (1976) The City as a Growth Machine: Toward a Political Economy of Place. *American Journal of Sociology,* 82(2), 309–332, www.jstor.org/stable/2777096

Montgomery, C. (2015) *Happy City: Transforming Our Lives Through Urban Design,* Penguin: London.

Moor, M. & Rowlnd, J. (eds.) (2006) *Urban Design Futures,* Routledge: London.

Moreno, C. (2021) *Definition of the 15-minute city: What Is The 15-minute City?,* Paper presented to Obel Award Jury, October, www.researchgate.net/publication/362839186

Moreno, C. (2024) *The 15-Minute City: A Solution to Saving Our Time and Our Planet,* Wiley: New York.

Morris, A.E.J. (1994) *History of Urban Form: Before the Industrial Revolutions,* Third Edition, Longman Scientific & Technical: New York.

Morris, C. (1971) *Writings on the General Theory of Signs,* Mouton de Gruyter: Paris.

Moskowitz, P. (2016) Bulldoze Jane Jacobs, *Slate,* 4 May, https://slate.com/business/2016/05/happy-100th-birthday-jane-jacobs-its-time-to-stop-deifying-you.html

Mossabir, R., Milligan, C. & Froggatt, K. (2021) Therapeutic Landscape Experiences of Everyday Geographies within the Wider Community: A Scoping Review, *Social Science & Medicine,* 279, https://doi.org/10.1016/j.socscimed.2021.113980.

Mostafavi, M. & Doherty, G. (eds.) (2010) *Ecological Urbanism,* Lars Muller Publishers: Zurich.

Moughtin, C. (2003) *Urban Design: Street and Square,* Third Edition, Architectural Press: Oxford.

Mumford, E. (2018) *Designing the Modern City: Urbanism Since 1850,* Yale University Press: New Haven.

Mumford, L. (1961) *The City in History: Its Origins, Its Transformations, and Its Prospects,* Harcourt, Brace & World, Inc.: New York.

Murphy, C. (2024) *The Mixed-Use Neighborhood: Creating a Sense of Place,* Wheatmark: Arizona.

Natarajan, L. & Short, M. (2023) *Engaged Urban Pedagogy: Participatory Practices in Planning and Place-Making,* UCL Press: London.

National Association of City Transport Officials (2013) *Urban Design Street Guide,* Island Press: New York.

Nelson, A.C. & Duncan, J. (1995) *Growth Management Principles and Practices.* Routledge: London.

Netek, R., Burian, J., Paszto, V., Barvír, R. & Chloupek, J. (2022) Two Decades of 'Brain Drain' in Olomouc (Czechia). *Journal of Maps*, 18(1), 125–132. https://doi.org/10.1080/17445647.2022.2099315

Newman, P. & Kenworthy, J. (2015) *The End of Automobile Dependence: How Cities are Mowing Beyond Car-Based Planning*, Island Press: Washington DC.

Norberg-Schulz, C. (1963) *Intentions in Architecture*, Allen & Unwin: London.

Norberg-Schulz, C. (1971) *Existence, Space and Architecture*, Praeger: New York.

Norberg-Schulz, C. (1979) *Genius Loci: Towards a Phenomenology of Architecture*, Rizzoli: New York.

Norberg-Schulz, C. (1988) *Architecture, Meaning and Place: Selected Essays*, Rizzoli: New York.

Oc, T. & Tiesdell, S. (1997) *Safer City Centres: Reviving the Public Realm*, Chapman: London.

OECD (2010) *Globalisation, Transport and the Environment*, OECD Publishing: Paris, https://doi.org/10.1787/9789264072916-en

OECD (2012) *Compact City Policies: A Comparative Assessment, OECD Green Growth Studies, The Organisation for Economic Co-operation and Development*, OECD Publishing, www.oecd.org/content/dam/oecd/en/publications/reports/2012/05/compact-city-policies_g1g191f1/9789264167865-en.pdf

Ohm, R.M. (1988) The Continuing Legacy of the Chicago School, *Sociological Perspectives*, 31(3), 360–376. https://doi.org/10.2307/1389204

Oldenburg, R. (1989) *The Great Good Place: Cafés, Coffee Shops, Community Centers, Beauty Parlors, General Stores, Bars, Hangouts, and how They Get You Through the Day*, Paragon House: New York.

Owens, E.J. (1992) *The City in the Greek and Roman World*, Routledge: London.

Page, M. & Mennel, T. (eds) (2011) *Reconsidering Jane Jacobs*, Planners Press: Chicago.

Palsson, K. (2023) *Urban Block Cities: 10 Design Principles for Contemporary Planning*, DOM: Freiburg.

Park, R.E. & Burgess, E.W. (1925) *The City*, University of Chicago Press: Chicago.

Perry, C. (1929) The Neighborhood Unit: A Scheme of Arrangement for the Family-Life Community. In *A Regional Plan for New York and Environs*, Vol. vii, New York.

Pineo, H. (2022) Towards Healthy Urbanism: Inclusive, Equitable and Sustainable (THRIVES) – An Urban Design and Planning Framework from Theory to Praxis, *Cities and Health*, 6(5), 9974–992, https://doi.org/10.1080/23748834.2020.1769527

Porta, S. & Latora, V. (2008) Centrality and Cities: Multiple Centrality Assessment as a Tool for Urban Analysis and Design. In T. Haas (ed.) *New Urbanism and Beyond: Designing Cities for the Future*, pp. 140–145. New York: Rizzoli International.

Portugali, J. (2024) Bohm's Theory of Orders as a Basis for a Unified Urban Theory, *Dialogues in Urban Research*, 2(3): 267–289. https://doi.org/10.1177/27541258241256972

Pradhan, P., Subedi, D.R., Dahal, K., Hu, Y., Gurung, P., Pokharel, S., Kafle, S., Khatri, B., Basyal, S., Gurung, M. & Joshi, A. (2024) Urban Agriculture Matters for Sustainable Development, *Cell Reports Sustainability*, 1(9), https://doi.org/10.1016/j.crsus.2024.100217

Pradhan, R.P., Arvin M.B. & Nair, M. (2021) Urbanization, Transportation Infrastructure, ICT, and Economic Growth: A Temporal Causal Analysis, *Cities*, 115, https://doi.org/10.1016/j.cities.2021.103213

Praharaj, S. & Han, H. (2019) Cutting Through the Clutter of Smart City Definitions: A Reading into the Smart City Perceptions in India, *City, Culture and Society*, 18, 100289, https://doi.org/10.1016/j.ccs.2019.05.005

Project for Public Spaces (2005) *What Makes a Successful Place?: The Place Diagram*, www.pps.org/article/grplacefeat

Project for Public Spaces (2008a) *Our Approach to Mixed-Use*, December, www.pps.org/article/mixeduseapproach (accessed 27th July 2023).

Project for Public Spaces (2008b) *Qualities of a Great Street*, 30 April, www.pps.org/article/qualitiesofagreatstreet

Project for Public Spaces (2022) *Placemaking: What If We Built Our Cities Around Places?: A Placemaking Primer*, PPS: New York.

Project for Public Spaces (2023a) *What If We Built Our Cities Around Places?*, www.pps.org/category/placemaking (accessed 04 August 2023).

Project for Public Spaces (2023b) *What Makes a Successful Place?*, www.pps.org/article/grplacefeat (accessed 04 August 2023).

Project for Public Spaces (2023c) *Eleven Principles for Creating Great Community Places*, www.pps.org/article/11steps (accessed 04 August 2023).

Project for Public Spaces (2023d) *The Power of 10+*, www.pps.org/article/the-power-of-10 (accessed 04 August 2023).

Project for Public Spaces (2024) *William H. Whyte: Placemaking Heroes*, www.pps.org/article/wwhyte

Punter, J. (ed.) (2010) *Urban Design and the British Urban Renaissance*, Routledge: London.

Punter, J. & Carmona, M. (1997) *The Design Dimension of Planning: Theory, Content, and Best Practice for Design Policies*, Spon Press: London.

Raco, M. (2009) Urban Governance, In Kitchen, R. & Thrift, N. (eds.), *International Encyclopedia of Human Geography*, pp. 622–627, Elsevier: London, https://doi.org/10.1016/B978-008044910-4.01089-0

Ramadhani, I.S. & Indradjati, P.N. (2023) Toward Contemporary City Branding in the Digital Era: Conceptualizing the Acceptability of City Branding on Social Media, *Open House International*, 48(4), 666–682. https://doi.org/10.1108/OHI-08-2022-0213

Randolph, G.F., & Storper, M. (2023) Is Urbanisation in the Global South Fundamentally Different? Comparative Global Urban Analysis for the 21st Century. *Urban Studies*, 60(1), 3–25. https://doi.org/10.1177/0042098021 1067926

Rein, R.K. (2022) *American Urbanist: How William H. Whyte's Unconventional Wisdom Reshaped Public Life*, Island Press: Washington DC.

Reynolds, R. & Reynolds, K. (2022) *Xenakis Creates in Architecture and Music: The Reynolds Desert House*, Routledge: London.

RICS (2021) *Urban Density: Promoting Sustainable Development*, Royal Institute of Chartered Surveyors, www.rics.org/news-insights/wbef/urban-density-promoting-sustainable-development-part-1 (accessed 18 November 2024).

Ritchie, A. & Thomas, R. (2009) *Sustainable Urban Design: An Environmental Approach*, Routledge: London.

Roberts, M. & Lloyd-Jones, T. (1997) Mixed Uses and Urban Design, In Coupland, A. (ed) *Reclaiming the City: Mixed Use Development*, pp. 149–178, Spon Press: London.

Rode, P. (2017) Urban Planning and Transport Policy Integration: The Role of Governance Hierarchies and Networks in London and Berlin, *Journal of Urban Affairs*, 41(1): 39–63. https://doi.org/10.1080/07352166.2016.1271663

Roe, J. & McCay, L. (2021) *Restorative Cities: Urban Design for Mental Health and Wellbeing*, Bloomsbury: London.

Rousset, I. (2021) The Industrious, the Laboring, and the Sunken: Berlin's Mietskaserne and the Housing Question, *Journal of Urban History*, 47(6), 1275–1300. https://doi.org/10.1177/0096144220917473

Rowe, P.G., van den Berg, H.J. & Wang, L. (2019) *Urban Blocks: History, Technical Features, and Outcomes*, Scholars' Press: Beau-Bassin.

RTPI (2014) *Promoting Healthy Cities: Why Planning is Critical to a Healthy Urban Future*, Royal Town Planning Institute, Planning Horizons No.3, October, www.rtpi.org.uk/media/1470/promoting-healthy-cities-full-rep ort-2014.pdf (accessed 7th July 2024).

Rudlin, D. & Falk, N. (2010) *Sustainable Urban Neighbourhood: Building the 21st Century Home*, Architectural Press: Oxford.

Ruming, K.J. (2014) Urban Consolidation, Strategic Planning and Community Opposition in Sydney, Australia: Unpacking Policy Knowledge and Public

Perceptions, *Land Use Policy*, 39, 254–265, https://doi.org/10.1016/j.landusepol.2014.02.010

Sadik-Khan, J. & Solomonow, S. (2017) *Street Fight: Handbook for an Urban Revolution*, Penguin Books: New York.

Salerno, R. (2014) *Rethinking Kevin Lynch's Lesson in Mapping Today's City*, In Contin, A., Paolini, P. & Salerno, R. (eds.) *Innovative Technologies in Urban Mapping: Built Space and Mental Space*, pp. 25–31, Springer: New York.

Sassen, S. (2001) *The Global City: New York, London, Tokyo*, Princeton University Press. https://doi.org/10.2307/j.ctt2jc93q

Savours, B. (2024) *The Planning Premium: The Value of Well-made Places*, June, Public First Limited: London, www.publicfirst.co.uk/wp-content/uploads/2024/06/The-Economic-Value-of-Good-Town-Planning-Final-1-1.pdf (accessed 10 December 2024).

Schlüter, J. & Weyer, J. (2019) Car Sharing as a Means to Raise Acceptance of Electric Vehicles: An Empirical Study on Regime Change in Automobility, *Transportation Research Part F: Traffic Psychology and Behaviour*, 60: 185–201, https://doi.org/10.1016/j.trf.2018.09.005

Schmid, C. (2022) *Henri Lefebvre and the Theory of the Production of Space*, Verso Books: New York.

Schmid, H. (2009) *Economy of Fascination: Dubai and Las Vegas as Themed Urban Landscapes*. Berlin: Gebr. Borntraeger.

Schubert, D. (ed) (2014) *Contemporary Perspectives on Jane Jacobs: Reassessing the Impacts of an Urban Visionary*, Ashgate: Farnham.

Shaftoe, H. (2008) *Convivial Urban Spaces: Creating Effective Public Places*, Earthscan: London.

Shao, Z., Sumari, N.S., Portnov, A., Ujoh, F., Musakwa, W. & Mandela, P.J. (2020) Urban sprawl and Its Impact on Sustainable Urban Development: A Combination of Remote Sensing and Social Media Data. *Geo Spatial Information Science*, 24(2), 241–255. https://doi.org/10.1080/10095020.2020.1787800

Shaw, B.J. & Houghton, D.S. (1991) Urban Consolidation: Beyond the Rhetoric. *Urban Policy and Research*, 9(2), 85–91. https://doi.org/10.1080/08111149108551463.

Sim, D. (2019) *Soft City: Building Density for Everyday Life*, Island Press: Washington.

Sitte, C. (2013) *The Art of Building Cities: City Building According to Its Artistic Principles*, Reprint of 1945 edition, Martino Fine Books: Connecticut.

Skelton, T. (2017) *Everyday Geographies*, Wiley: London.

Soja, E.W. (1996) *Thirdspace: Journeys to Los Angeles and Other Real-and-Imagined Places*, Wiley-Blackwell: Hoboken.

Soltani, A., Pieters, J., Young, J. & Sun, Z. (2017) Exploring city branding strategies and their impacts on local tourism success, the case study of

Kumamoto Prefecture, Japan, *Asia Pacific Journal of Tourism Research*, 23(2), 158–169. https://doi.org/10.1080/10941665.2017.1410195

Song, K., Chen, Y., Duan, Y. & Zheng, Y. (2023) Urban governance: A review of intellectual structure and topic evolution, *Urban Governance*, 3(3), 169–185, https://doi.org/10.1016/j.ugj.2023.06.001

Speck, J. (2013) *Walkable City: How Downtown Can Save America One Step at a Time*, North Point Press: New York.

Speck, J. (2019) *Walkable City Rules: 101 Steps to Making Better Places*, Island Press: Washington DC.

Spreiregen, P.D. (1965) *Urban Design: The Architecture of Towns and Cities*, McGraw Hill: New York.

Sullivan, J.D., Rogers, J. & Bettcher, K.E. (2007) The Importance of Property Rights to Development, *The SAIS Review of International Affairs*, 27(2), 31–43, www.jstor.org/stable/27000088

Tarbatt, J. & Street-Tarbatt, C. (2020) *The Urban Block: A Guide for Urban Designers, Architects and Town Planners*, RIBA Publishing: London.

Taylor P.J. & Derudder B. (2015) *World City Network – A Global Urban Analysis*. Routledge: New York.

Theodore D. (2016) Better Design, Better Hospitals, *CMAJ*, 188(12): 902–903, https://doi.org/10.1503/cmaj.151228

Thomas, D. (2017) *Placemaking: An Urban Design Methodology*, Routledge: London.

Thwaites, K., Mathers, A. & Simkins, I. (2013) *Socially Restorative Urbanism: The Theory, Process, and Practice of Experiemics*, Routledge: London.

Tiesdell, S. & Carmona, M. (eds) (2007) *Urban Design Reader*, Routledge: Oxford.

Tiesdell, S., Oc, T. & Heath, T. (1996) *Revitalizing Historic Urban Quarters*, Architectural Press: Oxford.

Tonkiss, F. (2013) *Cities by Design: The Social Life of Urban Form*, Polity Press: Cambridge.

Toti, A. & Yang, Z. (eds) (2019) *Grids of Chinese Ancient Cities: Spatial Planning Tools for Achieving Social Aims*, Altralinea Edizoni: Firenze.

Trancik, R. (1986) *Finding Lost Space: Theories of Urban Design*, John Wiley: New York.

Transport for London (2010) *The Mayor's Transport Strategy*, Greater London Authority: London.

Tyrnauer, M. (2017) *Citizen Jane: Battle for the City*, Altimeter Films: New York.

UKGBC (n.d.) *Climate Change Mitigation*, UK Green Building Council, https://ukgbc.org/our-work/climate-change-mitigation/#:~:text=Our%20planet%20is%20in%20crisis,to%20net%20zero%20by%202050 (accessed 7th July 2023).

UN Environment Programme (2024) *Cities and Climate Change,* www.unep.org/explore-topics/resource-efficiency/what-we-do/cities-and-climate-change (accessed 14th March 2024).

UNESCO (2016) *Culture Urban Future: Global Report for on Culture for Sustainable Urban Development,* UNESCO: Paris.

United Nations (2024) *Around 2.5 Billion More People Will Be Living in Cities by 2050,* www.un.org/en/desa/around-25-billion-more-people-will-be-living-cities-2050-projects-new-un-report (accessed 29th May 2025).

United Nations (2025) *Global Issues: Population,* www.un.org/en/global-issues/population (accessed 4th January 2025).

Urban Design Associates (2013) *The Urban Design Handbook: Techniques and Working Methods,* W.W. Norton: New York.

Urban Design Group (2024) *History of the Urban Design Group: Forty Years of the Urban Design Group,* www.udg.org.uk/about/history (accessed 23rd June 2024).

Urban Design Group (2025) *What is Urban Design?,* www.udg.org.uk/about/what-is-urban-design (accessed 15th October 2024).

Urban Design Lab (2025) *Definitions of Urban Design,* https://urbandesignlab.in/definitions-of-urban-design (accessed 12 December 2024).

Urban Design London (2017) *The Design Companion for Planning and Placemaking,* RIBA Publishing: London.

Urban Land Institute (2019) *Understanding Mixed Use and Multi Use,* https://knowledge.uli.org/-/media/files/reading-list/reading-list-pdfs/readinglist_mixeduse_v1.pdf?rev=cf67335ad1b44772b6f20bb074043a15 (accessed 23rd June 2023).

Urban Land Institute & Centre for Liveable Cities (2013) *10 Principles for Liveable High Density Cities. Lessons from Singapore,* Centre for Liveable Cities: Singapore, https://isomer-user-content.by.gov.sg/50/39ec0988-e0be-4777-96ee-7dea5f402c4f/10principlesforliveablehighdensitycitieslessonsfromsingapore.pdf (accessed 10th January 2025).

Urban Task Force (1999) *Towards an Urban Renaissance,* Routledge: London.

urbanNext (2025) *Flexible Urbanisms: Towards an Incremental Urbanism,* https://urbannext.net/flexible-urbanisms/ (accessed 14 January 2025).

van Raaij, W.F. (1993) Postmodern Consumption, *Journal of Economic Psychology,* 14(3),: 541–563, https://doi.org/10.1016/0167-4870(93)90032-G

Varna, G. and Tiesdell, G. (2010) Assessing the Publicness of Public Space: The Star Model of Publicness, *Journal of Urban Design,* 15(4), 575–598. https://doi.org/10.1080/13574809.2010.502350

Vazquez, F., Millard-Ball, A. & Barrington-Leigh, C. (2023) Urban Development and Street-Network Sprawl in Tokyo, *Journal of Urbanism: International Research on Placemaking and Urban Sustainability,* 1–20. https://doi.org/10.1080/17549175.2023.2262698

Verkade, T. & te Brömmelstroet, M. (2022) *Movement: How to Take Back Our Streets and Transform Our Lives,* Scribe Publications: London.

Verstegen, I. & Ceen, A. (eds) (2013) *Giambattista Nolli and Rome: Mapping the City before and after the Pianta Grande,* Stadium Urbis: Rome.

Wachten, K. & Neubauer, H. (2010) *Urban Design and Architecture: The 20th Century,* H.F. Ullmann Publishing: Cologne.

Wang, Y., & Wiedmann, F. (2024) Everyday Urbanism in Beijing's Edge Cities: On Spatial and Experience Patterns. *Journal of Urbanism: International Research on Placemaking and Urban Sustainability,* 1–21. https://doi.org/10.1080/17549175.2024.2383928

Ward, S. (2011) *The Garden City: Past, Present and Future,* Spon Press: London.

Ward, S. (2016) *The Peaceful Path: Building Garden Cities and New Towns,* University of Hertfordshire Press: Hatfield.

Weaver, T. (2017) Urban Crisis: The Genealogy of a Concept, *Urban Studies,* 54(9), 2039–2055, www.jstor.org/stable/26151464

Weber, M. (1930) *The Protestant Ethic and the Spirit of Capitalism,* Scribner/Simon & Schuster: London.

Welter, V.M. & Lawson, J. (eds) (2000) *The City After Patrick Geddes,* Verlag Peter Lang: Oxford.

Wheeler, S.M. (2023) *The Sustainable Urban Development Reader,* Routledge: London.

Whittle, N. (2021) *The 15-Minute City: Global Change Through Local Living,* Luath Press: Edinburgh.

Whittle, N. (2024) *Shrink the City: The 15-Minute City Urban Experiment and the Cities of the Future,* The Experiment: New York.

Whyte, W.H. (1988) *City: Rediscovering the Center,* University of Pennsylvania Press: Philadelphia.

Whyte, W.H. (2021) *The Social Life of Small Urban Spaces,* 8th edition, Project for Public Spaces: New York.

Wiedmann, F. (2013) The Verticalization of Manama's Urban Periphery, *Open House International,* 38(4): 90–100, https://doi.org/10.1108/OHI-04-2013-B0010

Wiedmann, F. & Salama, A.M. (2019a) *Building Migrant Cities in the Gulf: Urban Transformations in the Middle East,* I.B. Taurus: London.

Wiedmann, F. & Salama, A.M. (2019b) Mapping Lefebvre's Theory on the Production of Space to an Integrated Approach for Sustainable Urbanism, In: Leary-Owhin, M.E. & McCarthy, J.P. (eds.) *The Routledge Handbook of Henri Lefebvre, The City and Urban Society,* Routledge: London.

Wiedmann, F., Salama, A.M. & Thierstein, A. (2012) A Framework for Investigating Urban Qualities in Emerging Knowledge Economies: The Case of Doha, *International Journal of Architectural Research,* 6(1), 42–56.

Wiedmann, F., Thomas, S. and Peterek, M. (2022) Economic Competitiveness. In: B. Giddings & R.J. Rogerson (eds.), *The Future of the City Centre: Global Perspectives*, Routledge: London.

Williams, J.M. (2016) Evaluating Mega Projects: The Case of Forest City in Johor, Malaysia. Malaysia Sustainable Cities Program, Working Paper Series, Accessible online (Boston: MIT): https://scienceimpact.mit.edu/sites/default/files/documents/Williams.pdf

Wirth, L. (1938) Urbanism as a Way of Life, *American Journal of Sociology*, 44(1): 1–24.

World Bank (2023) *Urban Development*, www.worldbank.org/en/topic/urban development/overview (accessed 15th June 2024).

Wray, I. (2016) Milton Keynes: The Making of a New City. *Geography*, 101(3), 116–124, www.jstor.org/stable/26546731

Wrigley, E.A. (1985) Urban Growth and Agricultural Change: England and the Continent in the Early Modern Period. *Journal of Interdisciplinary History*, 15(4), 683–728.

Wunderlich, F.M. (2023) *Temporal Urban Design: Temporality, Rhythm and Place*, Routledge: London.

Wyckoff, A. (2014) Definition of Placemaking: Four Different Types. *Planning & Zoning News*, www.canr.msu.edu/uploads/375/65814/4typesplacemaking_pzn_wyckoff_january2014.pdf

Xie, J. & Heath, T. (2018) *Heritage-led Urban Regeneration in China*, Routledge: New York.

Xue, E.Y. (2022) Tourism as Creative Destruction: Place Making and Resilience in Rural Areas, *Journal of Tourism and Cultural Change*, 20(6), 827–841. https://doi.org/10.1080/14766825.2022.2114359

Young, R. & Clavel, P. (2017) Planning Living Cities: Patrick Geddes' Legacy in the New Millennium, *Landscape and Urban Planning*, 166(Special Issue), www.sciencedirect.com/journal/landscape-and-urban-planning/vol/166/suppl/C

Zhang, X. & He, Y. (2020) What Makes Public Space Public? The Chaos of Public Space Definitions and a New Epistemological Approach. *Administration & Society*, 52(5), 749–770. https://doi.org/10.1177/0095399719852897

Zucker, P. (1959) *Town and Square: From the Agora to the Village Green*, Columbia University Press: New York.

Zukin, S. (2010) *Naked City: The Death and Life of Authentic Urban Places*, Oxford University Press: New York.

INDEX

access 3, 8, 28, 31, 35–9, 61, 65, 67, 70, 90, 96, 102, 107, 111, 119–20, 131, 134, 158–9, 161–6, 174–5, 179–81, 189–91, 196, 202–3, 205, 209, 213
active edges 48, 67, 81–2, 88, 91, 93–4, 107, 130, 171
adaptability 45, 57, 61, 101, 116, 140
adaptive re-use 100, 102, 110
Africa 21, 85
Alexander, Christopher 56–7, 100
American Institute of Architecture 6
Amsterdam 35, 129, 191
Asia 16, 28, 60, 190
Athens 17
Athens Charter 30
Australia 28, 60, 185

Bacon, Edmond 5
Bahrain 162–3
Bangkok 85
Barcelona 19, 39, 81, 85, 165, 192
Baroque 23
Battista Alberti, Leon 22
Beijing 5, 66, 84, 86, 121, 195
Berlin 192–3
Bernini, Gianlorenzo 23

Boston 77
Brasilia 31, 189
Broadacre City 31–2
brownfield development 100, 188, 195–6
Burgess, Ernest 145, 147–8
Burnham, Daniel 26–7

Cairo 85, 190
cars 3, 5, 31, 33–4, 38, 56, 90–1, 106, 110–11, 115–16, 130, 150, 155, 157, 159, 166, 174, 184–9, 191, 194, 198–9
Castells, Manuel 153–4
Cerdà, Ildefons 19, 192
Chandigarh 31
character 3, 36, 47, 51, 53, 55, 69–71, 73, 81, 101, 107, 109
Chicago 19, 26–7, 66, 146, 147
Chicago School of Sociology 13, 141, 143, 147–8, 150
China 4–5, 16, 18, 37, 66, 86, 93, 106, 109, 120–1, 125–6, 128, 195, 213
CIAM 30–2
City Beautiful 15, 26–7
Cleveland 27
climate 21, 68, 74, 126–7, 165, 201, 209

climate change 1–3, 124
comfort 21, 36, 55, 61, 67–8, 70, 115–16, 119, 122–3, 127, 130, 152, 171–2, 186, 190, 212
communities 3, 6–7, 9, 16, 25, 27–30, 32, 34–6, 38–9, 42, 45–7, 49, 54, 60–2, 64, 68, 76, 79, 83, 85, 87, 91, 100, 104, 108, 115, 122–3, 131, 133–4, 136–7, 145–6, 172, 179, 181–2, 185, 188–9, 191, 201–2, 204, 213–14
community 7–8, 12, 28–9, 33, 35–40, 45–8, 55, 61, 78, 89, 102, 104, 108–9, 119, 131, 133, 146, 173, 192, 205, 211
commuting 114, 166, 174, 181, 194, 210
compact city 34–5, 37–8, 90
competitions 203–6
Congress for New Urbanism (CNU) 35–6
connections 3, 8, 15, 34, 41, 69, 78, 148, 154, 166, 171, 186–7, 199, 201–2, 207, 213
connectivity 76, 88, 96, 123, 179, 187, 202
congestion 3, 6, 27, 107, 111, 139, 184–5, 200
context 4, 13, 20, 32, 39–40, 47, 55, 61, 64–6, 69, 82–3, 85–9, 119, 123, 129, 131, 139–40, 142–8, 150, 154, 156, 160, 162–3, 167–9, 171, 176–7, 179–81, 183, 194, 201, 204–5, 209, 214
contextual 10, 35, 63, 85, 88–9, 140
Copenhagen 60, 67, 118, 196
crime 68, 129–30
Cullen, Gordon 41, 50–1
cycling 3, 34–6, 38, 49, 70–1, 96, 106–7, 110–12, 115–16, 118, 130, 138, 166, 184

decision-making 1, 13, 47, 59, 104, 131, 136, 147–9, 151–2, 156, 176, 178, 180–3, 187–8, 204, 208
density 31–5, 37, 44, 88, 90, 145–6, 161, 186, 192
developers 9, 12, 35, 182, 187, 194, 202–4, 206
diversification 13, 160–7, 183
Doha 85
Duany, Andres 36–7
Dubai 132, 169

economic 3, 7–10, 20, 24, 27, 34, 36, 43, 47–9, 65, 69, 73–4, 83, 89, 91, 94, 101, 103, 105–6, 108–9, 111, 115, 117, 119, 122, 131, 134, 137–9, 143–4, 153–4, 156–60, 163–4, 167, 173–4, 177–81, 183–4, 187, 189–90, 192, 195–6, 200–2, 205–7, 209–10, 212–13
Edinburgh 19, 44–5
Egypt 16
enclosure 43, 69, 71, 76, 79–80
energy 107, 110, 158, 160, 174, 185
engagement 10–11, 46–7, 49, 54, 73, 134, 136–7, 173, 178, 187, 212
English Partnerships 201
environment 2–5, 7, 9–12, 20–1, 28–9, 34–6, 39, 42, 44–7, 49–3, 55, 57–9, 64–6, 69–70, 74, 76, 81–2, 85, 87, 90, 92, 94, 96, 98, 102, 111–14, 119, 121, 123–4, 126–7, 131, 135, 138, 140–3, 145, 149, 154–5, 158, 160, 168, 171–3, 182, 186, 190–1, 202–3, 210, 212
environmental 7, 9, 20–1, 27, 36, 49, 52, 68, 70, 87, 89, 110, 124, 139, 141, 143–4, 152, 155, 157–8, 167, 205, 207
Essex Design Guide 68, 90
Europe 2, 17, 20–1, 23–6, 28, 30, 32, 34–5, 43, 60, 71, 81, 84–5, 151, 185, 192–4, 198

Filarete 22 and 15-minute city 35, 37–9, 107, 111, 213
flexible 57, 61, 74, 81, 98–9, 123, 162, 165–6, 191, 213
flexibility 78–9, 88, 99, 104
Florence 84–5
fragmentation 174, 183, 185, 209
Frankfurt am Main 149, 194, 198

Garden City 26–8, 33, 38, 44, 188
Geddes, Patrick 41, 44–5, 58
Gehl, Jan 42, 58–60, 68, 119, 122
genius loci 41, 52, 54–5, 177
geography 10–11, 141, 143–4, 167
Glasgow 165, 199
globalisation 85, 151, 156, 158, 207, 210
Global South 179
governance 13, 142, 144–5, 175–83, 185, 187, 191, 201–2, 207–8
Greece 17–18
Greek Empire 15, 17
greenfield development 188, 191–2, 194–6, 201
Greenwich Village 46
grids 16–19, 22, 26, 31, 161, 164–6, 187, 190, 192, 194–5
growth 1–2, 13, 15, 25, 28, 33–6, 40, 44, 48, 88, 90, 100, 117, 138, 145, 147–8, 156–9, 162, 174, 178–81, 183–5, 187–8, 190–1, 194, 196, 200, 206–8, 214

Hafen City, Hamburg 196
Hall, Peter 154
Harvard University 5–6
Haussmann, Baron Georges-Eugene 25–6
health 2, 18–19, 26, 28, 36, 38–9, 45, 70, 81, 102, 107–8, 111, 115, 119, 124, 126, 137, 139, 155, 171, 209–14

heritage 49, 89, 109, 162–4, 172, 197–8, 205
High Line 101
Hippodamus of Miletus 17
Hispaniola 18
Hong Kong 34, 179
housing 2, 25, 27, 34, 37–8, 45–7, 90, 102, 108, 142, 145, 148–9, 151, 157, 160–2, 165–6, 173–4, 179, 185, 188, 191–2, 194–5, 200, 209–10, 213
Howard, Ebenezer 27–8, 38, 44
human-centric 7, 12, 38, 40, 42, 63, 110, 119

ideal city 22–4
identity 20, 51–2, 55, 60, 67, 76, 89, 110, 131
inclusive 7, 9–11, 35, 38–40, 45, 47, 61, 66, 70, 90, 107, 112, 117, 119, 131, 136
incremental 20, 48, 88, 100–102
India 31, 45, 190
Indus River Valley 16
Industrial Revolution 24, 43, 147, 150, 184, 192
infill 36, 102
infrastructure 16, 18, 25, 28, 30, 44, 72, 76, 99–102, 111, 118, 122, 126, 134, 209–10, 214
isolation 3, 140, 174, 188, 212
Italy 2, 21–2, 34, 95

Jacobs, Jane 5, 27, 34, 38, 41, 46–50, 52, 69, 82, 90, 108, 162, 180, 210
Joint Centre for Urban Design 7

Kuala Lumpur 70, 85, 112, 127

landmarks 23, 41, 45, 48, 52, 78, 82, 97–8, 189, 193, 196

landscape 7, 10–11, 28–9, 31, 131, 134, 150, 168, 184–5, 187–9, 191, 195, 205, 209
landscape architecture 4–5, 7, 26
land-use 11, 19, 29, 32, 35, 44, 93, 103–4, 106–7, 115–16, 130, 150, 161, 173, 178, 181, 186–7, 190, 192, 194, 196, 203, 205
land-use zoning 3, 35, 42, 90, 116, 160–2, 192
Latin America 84
Law of the Indies 18
Le Corbusier 30–1
Lefebvre, Henri 13, 141, 144, 150–2, 154–6, 174
legibility 52, 81, 88, 97
L'Enfant, Pierre Charles 26
Letchworth Garden City 28
liverpool 196–7
Lloyd Wright, Frank 31
London 21, 25, 28, 32, 34, 76, 111, 147, 160–2, 165, 168, 179, 188, 198, 211
Lynch, Kevin 41, 51–4

Madrid 84
Malaysia 70, 112, 127, 190
management 3, 8, 10–11, 41, 61, 65, 67, 79, 87, 90, 127, 131, 134–6, 167, 174, 179–81, 185–6, 207
markets 16–17, 21–2, 25, 37, 65, 78, 84–6, 104, 134, 165, 167, 171, 186, 206, 212
Marx, Karl 150–1
masterplans 11, 18, 25, 28, 30, 36–7, 81, 187, 189–92, 194–5, 197, 201–4, 206
McKim, Charles 26
McMillan Commission 26
Mead, George H. 147
Mesopotamia 16
Mexico 19, 37

Michelangelo 22
micro-climate 126
Middle Ages 43
Middle East 71, 85, 190
Milton Keynes 19, 33, 188–9
mixed-use 34, 36–8, 42, 48, 102–3, 105, 107, 164, 173
mobility 69–70, 115, 131, 142, 146, 149, 161, 166, 173–4, 187
modernisation 156, 158, 178–80, 185, 199, 207
modernism 30, 32, 81, 103, 192
modernist city 30–1
mono-functional 3–4, 31, 42, 47, 103, 107, 164, 183
More, Sir Thomas 24
Moreno, Carlos 38–9
Moses, Robert 47
multi-modal 102, 112, 115, 117
Mumbai 190, 195
Mumford, Lewis 5, 20, 23
Munich 198

natural surveillance 3, 48, 80, 92–3, 95, 111, 115, 119, 129, 130
neighbourhoods 7, 12, 29, 37–9, 42, 46–9, 57, 67, 71, 82, 90, 100, 103, 106–10, 115, 122, 137, 147, 153–4, 167, 171, 173–4, 181, 183–7, 191–2, 200, 212
neighbourhood unit 29–30
New Urbanism 28, 35–7, 57
New York 19, 29, 46–7, 53, 168, 179
Nolli, Giambattista 41
Norberg-Schulz, Christian 54–5
North America 5–6, 25–8, 31, 34, 46, 51, 60, 185
Nottingham 9, 78, 83–4, 117, 136, 182

Olmsted, Frederick Law 26
orthogonal city planning 17–19

Palmanova 22
Paris 25–6, 31, 38–9, 71–2, 75–6, 81, 94, 99, 124, 135, 137, 151, 165, 191
Park, Robert 145, 147
participation 10, 20, 42, 45, 134, 136–7, 175, 190, 214
pedestrian 3, 5, 18, 33–7, 44, 47, 53, 56, 66–7, 93, 96, 107, 111, 115–17, 130, 166, 169, 184, 187, 192, 196, 198–9
people-friendly 60, 73
Perry, Clarence A. 29, 38
Philadelphia 19, 84
Piazza del Campidoglio 22
Piazza del Popolo 23, 76
Piazza Navona 23
place 1–2, 8–13, 16, 20–1, 24, 28, 30, 36, 41–2, 45, 47–9, 51–3, 55–69, 71, 73–6, 78–9, 82, 84–9, 91, 96, 99–102, 104–11, 114–15, 119–20, 122–5, 127–31, 133–40, 148, 151–3, 160, 167–74, 176–7, 183, 185, 189, 193–4, 197, 199, 201, 205–7, 209–10, 212, 214
place attachment 67, 100, 152–3, 167, 172–3, 206–7, 210
Place Diagram 61, 67
placemaking 7, 13, 41–2, 51, 61–3, 65–6, 123, 172, 177, 193, 201, 210, 212, 214
Plater-Zyberk, Elizabeth 36
policy 6, 8, 11, 38, 53, 148, 163, 198, 202
policy makers 39, 188, 201, 204
population 1–2, 24, 27–8, 31, 33, 37, 90, 111, 145–6, 154, 159, 171–2, 190, 210
population growth 1–2, 33, 90, 148, 158–9
Poundbury 35, 37
Power of 10+ 61, 104

private space 41, 48, 81, 137, 159, 165, 184, 189
privatisation 159, 209
Project for Public Spaces (PPS) 42, 54, 60–2, 66–7, 103–4
Pruitt-Igoe Estate 32
public realm 4, 8, 11, 14, 40–4, 50, 53–4, 59, 62–5, 68–9, 73, 80–1, 83, 85, 87–94, 96–8, 102–4, 111, 115, 119, 122–4, 126–7, 129–31, 133–4, 136–40, 163, 187, 189, 198, 207, 211
public transport 3, 27, 29, 33–6, 39, 48, 70, 90–1, 96, 109, 115, 117, 134, 138, 185

Qatar 190, 201

regeneration 46, 85, 197–200, 202
renaissance 15, 21–4, 27, 37, 43, 151, 193
renewal 6, 30, 34–5, 46–7, 161, 164, 188, 200
Riyadh 153–4, 186, 189
Roman Empire 17, 21, 23, 76, 80, 84
Rome 17–18, 22–3, 41

St. Peter's Basilica 23
safety 3, 8, 20, 47–8, 70, 92, 94, 96, 115, 119, 129, 152, 163, 165, 168–71, 184, 186, 212
Santiago 84
Saudi Arabia 153–4, 186, 189
seaside 36–7
seating 93–4, 111, 116, 123, 126
serial vision 50–1
Sert, Josep Lluís 5–6
settlements 3–4, 15–16, 18–22, 28, 31, 33–4, 36–7, 39–40, 50, 56, 64–5, 71, 90, 140, 145, 152–3, 165, 179, 185, 188, 191–2, 194
Sforzinda 22

Shanghai 39, 109, 120, 190
Sienna 2, 76, 84
Singapore 105, 179
Sitte, Camillo 43–4
smart city 35, 158–9, 176, 190, 213
social science 141–4, 150, 210
sociology 10, 11, 58, 141, 143–7, 149
de Soissons, Louis 28
Soja, Edward 152
South America 18, 28, 84
South-East Asia 71, 85, 190
Spain 18, 77, 79, 86, 97, 114, 116, 138
squares 9, 21–4, 26, 43, 47, 64–5, 67, 69, 74–81, 83–8, 91, 94, 96, 98, 104, 111, 119, 123, 126, 131, 134, 136–7, 139–40, 161, 186, 211
stakeholders 136–7, 177–8, 180, 187, 191, 202, 204, 206
standardisation 32, 85, 176
street furniture 112, 123, 130
streets 16–23, 25–6, 29, 33, 36–7, 42–5, 47–8, 53, 64–6, 69–76, 80–3, 85, 87–8, 91–6, 98, 102–3, 105, 111–12, 115–17, 119–21, 123, 126–7, 129–31, 134, 137–40, 153–4, 162, 164–6, 171, 173, 182, 184, 186–7, 193–4, 199, 202, 212, 214
sustainable 3, 8, 12–13, 35–9, 56, 61, 64, 70, 73, 90–1, 101–2, 105–7, 110–11, 115, 124, 139–41, 144, 156, 160, 166–7, 174, 176–7, 183, 187
sustainability 40, 63, 79, 102, 107–8, 110, 124, 133–4, 141, 159, 163

Taiwan 6, 84, 211
temporary activities 104, 132, 135
tenure 47–8, 102, 109

time 3, 9, 19, 25, 27–8, 31, 39, 41, 44–6, 50, 53, 56–8, 65, 78, 89, 98, 103–4, 110, 115, 120, 122, 130, 134, 138, 144–6, 148, 150, 152, 155–6, 158–9, 167, 170–1, 174, 177, 181, 184–5, 202–3, 210
Tokyo 85, 160
tourism 10, 103, 172, 183, 200
traffic 3, 6, 35–6, 67–8, 80, 96, 107, 111, 130, 162, 166, 171, 184–5, 187, 198–9, 205
transit-oriented 67, 96, 102, 112, 115–17, 126, 149, 160, 166, 185, 187, 194, 196
transport 3, 10–11, 25–7, 29–31, 33–6, 39, 44, 48, 69–70, 72, 76, 90–1, 96, 101, 111, 115, 117, 134, 138, 154, 184–7, 194, 198, 202, 210, 213

United Kingdom (UK) 7–9, 21, 24–8, 33, 35, 37, 73, 75, 78, 83, 100, 117, 121, 133, 136, 147, 161, 182, 188–9, 197–9, 201, 211
United Nations 141
United States (US) 5, 19, 26, 28, 31–2, 34–5, 37, 54, 62, 145, 147–8, 165, 169
urban block 11, 19, 34, 36, 47–8, 64, 69, 80–2, 88, 103, 140, 162, 164–5, 171, 187, 192–3
Urban Design Group 7–10, 12
urbanisation 1–2, 4, 16, 24, 31, 147, 156–7, 179, 188
urbanism 13, 15, 18, 20, 43, 81, 104, 141, 144–6, 148–54, 156, 159–60, 164, 167, 169, 172, 174, 176–7, 183–4, 188–9, 192, 201, 205, 207, 210
urban sprawl 3–4, 17, 32, 35, 90, 150–1, 181, 185, 192, 194, 196

Urban Task Force 37
users 8, 35, 48, 54–5, 57, 59, 61, 67–70, 78–9, 87–8, 91, 93–4, 103–4, 106, 115–16, 119–20, 122–3, 127, 129–32, 134–6, 139
Utopia 24, 31

vernacular 19–21, 30, 37, 152–3, 165, 172, 201
Vienna 32, 72
vitality 42, 47–9, 90, 103, 108–9, 119
Vitruvius 18, 22

walking 3, 34, 36, 38, 49, 58, 62, 68, 70–1, 93, 106–7, 110–12, 115, 138, 155, 165–6, 170–1, 193, 195, 210
walkability 33, 36, 48, 59, 91, 102, 110, 112, 115, 195
Washington D.C. 26, 72
waterfronts 65, 196–7
well-being 26, 39, 117, 127, 131, 139, 168–71, 208, 211
Whyte, William H. 41, 53–4, 60
Wirth, Louis 141, 144–9
World War II (WWII) 32–5, 37, 50, 150, 162, 181, 191

zoning 3, 35, 42, 90, 103, 116, 160–2, 192

For Product Safety Concerns and Information please contact our EU representative GPSR@taylorandfrancis.com
Taylor & Francis Verlag GmbH, Kaufingerstraße 24, 80331 München, Germany

www.ingramcontent.com/pod-product-compliance
Lightning Source LLC
Chambersburg PA
CBHW070839160426
43192CB00012B/2244